D0734328

A Pocket Guide to the Stars & Planets

A Pocket Guide to the
Stars & Planets

DUNCAN JOHN

Bath · New York · Singapore · Hong Kong · Cologne · Delhi · Melbourne

This is a Parragon Publishing Book
First published in 2006

Parragon Publishing
Queen Street House, 4 Queen Street, Bath, BA1 1HE, UK

Produced by Atlantic Publishing

Photographs courtesy of Science Photo Library

ISBN 978-1-4054-7337-8

Printed in China

CONTENTS

INTRODUCTION

The term astronomy derives from the Ancient Greek phrase meaning the laws or science governing the stars. However, astronomy is not just the study of the stars; rather it is the study of all celestial phenomena. It combines all the other Earth sciences and applies them to the study of all the matter existing outside our own planet, making it the widest ranging science. In short, astronomy is the science of understanding everything that goes on beyond Earth's boundaries. Astronomy is also one of the oldest of sciences. Every civilization, through antiquity to the recent past, had stunning views of the stars night after night, as sightings of the cosmos would not have been hampered by light pollution and an indoor lifestyle, both of which hide much of the heavens from observers today. All over the world some sort of understanding of the celestial sphere was an integral part of civilization, whether or not they opted for scientific explanations of the observed phenomena.

The Ancient Greek philosopher Ptolemy asserted that the Earth was at the center of the universe, and all the bodies in the sky orbited the Earth in eternal circles with perfect motion. This is known as the geocentric theory, as opposed to the heliocentric theory which places the sun at the center.

A hundred years earlier, in the third century B.C. Aristarchus had proposed a heliocentric theory of the universe, but this was never popular among the Greeks and Ptolemy's model found favor. This view influenced thought during the Middle Ages in Western Europe where the geocentric model, supported by the Church, became sacrosanct.

Nicolaus Copernicus was born in the Polish town of Torun in 1473 just as the Middle Ages were giving way to the Renaissance era, which saw an emphasis placed upon careful, scientific observation in astronomy. Copernicus believed that the Ptolemaic system did not properly account for what he observed in the heavens. Copernicus became the first to realize the true order of the planets in the solar system; Earth was demoted from the center of the universe to third rock from the sun.

This new wave of astronomical thinking did not arouse the interest of the Church too much until it reached the Italian peninsula at the beginning of the seventeenth century. Galileo Galilei was unwilling to commit to a heliocentric model until there was sufficient observational evidence to prove it. To gather such evidence, Galileo pioneered the use of a telescope in astronomy by advancing the new Dutch invention for studying the heavens.

400 years on, our knowledge of the solar system and what lies beyond may be considerable, yet the sense of awe is hardly less than that experienced by the earliest observers.

A Pocket Guide to the Stars and Planets takes us on a journey of discovery, from a detailed study of the planets in our solar system to the far-distant galaxies and outermost reaches of the universe. All the spectacular cosmic phenomena are featured, including comets, asteroids, meteors and nebulae. Some may be less familiar, such as quasars, dark matter, which forms some 90 percent of the universe even though it cannot be detected directly; and black holes, superdense matter with a gravitational pull so strong that not even light can escape.

The canvas of space and time is vast: 15 billion years since the Big Bang, when the entire universe occupied the space of a sub-atomic particle; 5 billion years since a cloud of gas and dust collapsed to form the star which sustains life on Earth. To study an object in deep space is to see hundreds of millions of years into the past, such are the mind-boggling distances that light has to travel.

Whether you're an inveterate star-gazer or a new student of astronomy, this lavishly illustrated book will enhance your knowledge and understanding of the universe. It includes the most recent images from the Hubble Telescope and Cassini probe, and thus provides an authoritative but accessible overview of the latest exciting developments in space research.

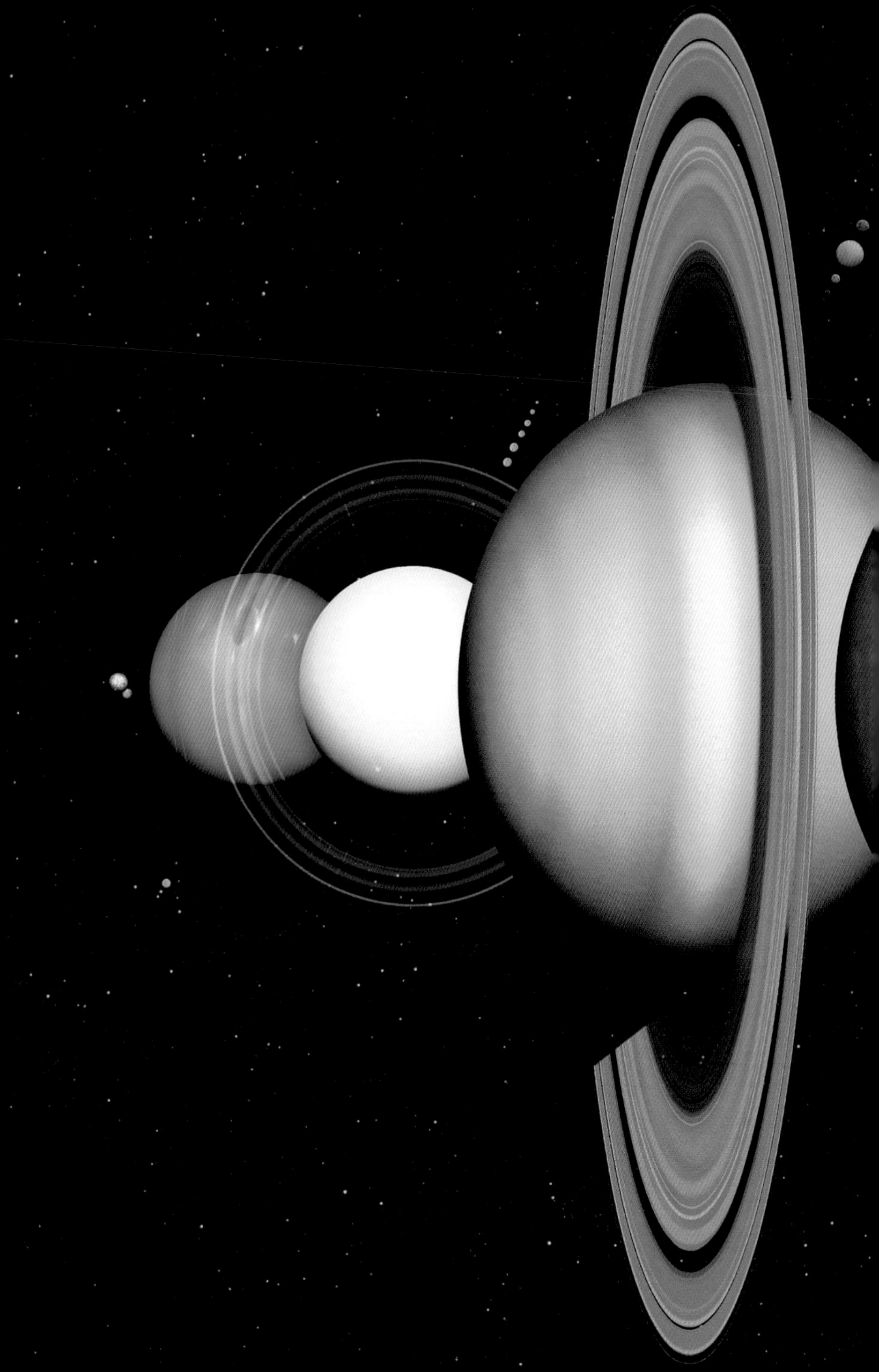

SOLAR SYSTEM

FORMATION

ABOVE: An amateur astronomer, silhouetted against the horizon, uses a refractor telescope to view Venus.

DISCOVERY OF THE PLANETS

Ancient discoveries Moon, Mercury, Venus, Mars, Jupiter, Saturn

1781	Uranus
1846	Neptune
1930	Pluto

The solar system is the name given to the planetary system of which the Earth is a part. It comprises planets, moons, comets, meteors and asteroids which are all held together by the gravitational pull of a star, named either the Sun or Sol. The solar system is believed to have formed from one nebula, the solar nebula. As gravity forced the nebula to condense it became more dense and pressure inside it increased, resulting in the creation of a proto-star, which began heating up to form the sun we see today. The proto-star would have been surrounded by interstellar dust and gases which began clumping together as a result of gravity. This process continued until, about 4.6 billion years ago, the clumps of rock and gas became much larger, and eventually gravity forced these irregular-shaped objects into the globular-shaped planets we see today. Many of the rocks did not become large enough to form planets and either remain today as asteroids, or they collided with the planets earlier in their history causing the large impact craters still visible throughout the solar system.

THE PLANETS

The known planets in the solar system can be divided into two groups. The four planets closest to the sun, Mercury, Venus, Earth and Mars, are called the "terrestrial planets" after the Latin word for "land" because they all share similar surfaces comprising solid rock surrounding dense, metallic cores. The four outer planets, Jupiter, Saturn, Uranus and Neptune are known as the "Jovian planets," implying their similarities to the planet Jupiter – they are all much larger than the terrestrial planets, and do not share their rocky surfaces and metallic core. Instead they are giant balls of atmosphere, mainly comprising gases surrounding relatively small rocky cores. The terrestrial and the Jovian planets are helpfully divided by a belt of asteroids, orbiting the sun between Mars and Jupiter. More recently Uranus and Neptune have been called Uranian rather than Jovian planets to highlight their differences to Jupiter and Saturn; mainly that they are appreciably smaller, are both bluish-greenish in color, comprise a significant amount of methane and have a thick coating of ice around their cores.

ABOVE: An illustration showing the relative sizes of the major bodies in the solar system. The sun is in lower frame. Superimposed upon it are the four inner (terrestrial) planets: Mercury, Venus, Earth (with the Moon to its left) and Mars. In the upper frame are the four major outer (gas giant) planets: Jupiter, Saturn, Uranus and Neptune. The furthest planet from the sun, Pluto, is seen upper right, along with its relatively large moon Charon.

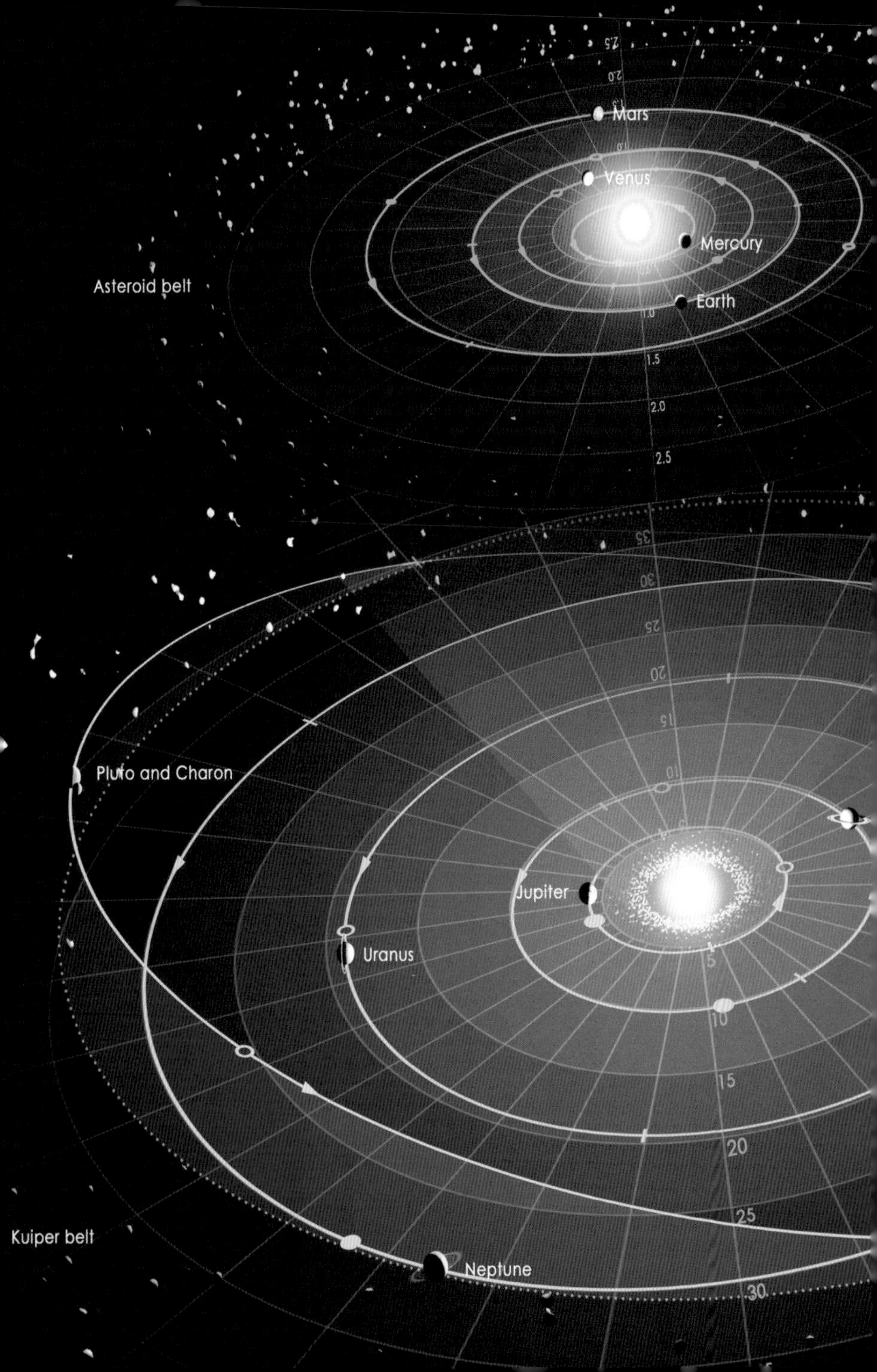

Mars
Venus
Mercury
Earth
Asteroid belt
1.0
1.5
2.0
2.5
Pluto and Charon
Jupiter
Uranus
Kuiper belt
Neptune
5
10
15
20
25
30

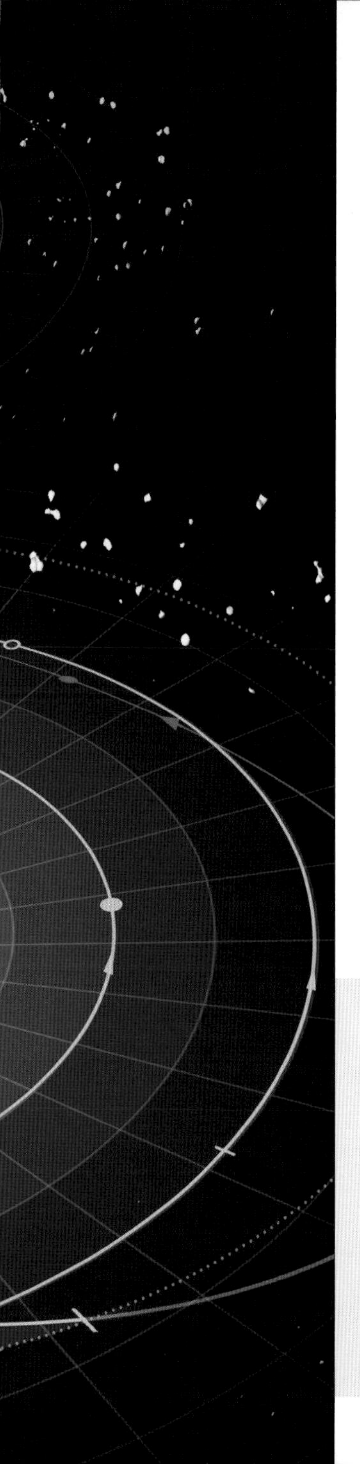

THE KUIPER BELT

Pluto stands out as somewhat anomalous in the solar system. As a tiny ball of rock and ice usually far beyond the orbit of Neptune, it does not fit easily into either terrestrial or Jovian category. Moreover, Pluto's orbit is highly eccentric, resembling that of a comet rather than a planet. It is also very small and, had it been discovered today, would probably not have been classified as a planet at all. Instead it is better seen as being one of a number of trans-Neptunian objects, planetoids, asteroids and centaurs which make up the Kuiper Belt.

The Kuiper Belt is a vast field of icy asteroids of varying sizes, centaurs and periodic comet nuclei orbiting the sun beyond Neptune. In July 2005, the discovery of a tenth planet orbiting the sun was announced. The anonymously named 2003 UB313 is larger than Pluto, and should be given planetary status if Pluto remains classified as a planet. However, improvements in telescopic technology have meant that more and more large objects from the Kuiper Belt have been discovered, such as Sedna and Quaoar. They are similar in size to Pluto, and although they are much farther out, some believe they should also gain planetary status, which would take the current number of planets to twelve.

LEFT: Computer artwork showing the planets of the solar system and their orbits. The terrestrial planets are separated from the gas planets. The asteroid belt (between Mars and Jupiter) and the Kuiper Belt (between and beyond Neptune and Pluto) are also shown. The open lines represent the points of orbit where the planets pass closest to the sun (perihelia), while the closed lines represent the points of orbit where the planets pass furthest from the sun (aphelia).

SOLAR SYSTEM

THE END OF THE SOLAR SYSTEM

It is difficult to calculate exactly where our solar system ends. It ends at the point at which objects are no longer affected by the sun's gravitational pull. This is also the area beyond the reach of solar winds and outside the sun's vast magnetosphere. The farthest reaches of the solar system are thought to be surrounded by a great halo, named the Oort Cloud, home to millions of comet nuclei and small icy rocks, which, it is speculated, lies a thousand times further from the sun than Pluto. However, more recent figures have suggested that Sedna might be a part of the Oort Cloud and not the Kuiper Belt, indicating that the cloud might be much closer to the sun than initially thought. If this is the case, estimates of the size of the solar system might need to be reduced. Voyagers 1 and 2 are the farthest-reaching man-made objects in the solar system. After completing successful missions to the outer planets during the 1980s, the two spacecraft continued towards the edge of the solar system with the intention of reaching its outer boundary and entering interstellar space.

OTHER PLANETARY SYSTEMS

In the early 1990s, Aleksander Wolszczan discovered the first four extra-solar planets orbiting the star PSR 1257+12. Extra-solar planets can be detected using two methods. The most popular is to look for "wobbles" – small movements of a star resulting from the gravitational influence of a planet. Movements of a star caused by even the largest of planets are so minor they are difficult to detect using telescopes. As such, only extra-solar planetary systems with Jupiter-sized planets in orbits very close to their suns have been easily detected using this method. Another method that is used is planetary transits. When Venus passes in front of the sun, its silhouette can be seen from Earth; likewise, when planets pass in front of a star, Earth-based telescopes note the dimming in the light of the star, which provides evidence of a planet around that star and can even show the planet's size. Again this process is only helpful for detecting Jupiter-sized planets, which cause sufficient dimming of the star to be detected on Earth, many light years away. NASA and the European Space Agency have plans to launch a Terrestrial Planet Finder in the coming decades, which will look for smaller, Earth-sized planets, and assess their chances of harboring life.

OPPOSITE: The optical telescopes of Keck I and II on Mauna Kea, an extinct volcano 4200 meters high. Comet Hale-Bopp, seen in early 1997, was one of the brightest comets of the 20th century. Here, its two tails are seen - the gas or "ion" tail and the dust tail.

COMET HALE-BOPP

1995 - Hale-Bopp was discovered by Alan Hale and Thomas Bopp and holds the record for the length of time it was visible to the naked eye (19 months)

Diameter is estimated to be about 40km

Rotation period = 11 hours 24 minutes

Length of tail = 80 - 95 million km

Distance (closest) from Earth = 196,000,000km

SUN

OUR STAR

ABOVE: Sunrise over the Moon. Sunlight on the Moon is much brighter than that on Earth, due to the Moon's lack of an atmosphere. The seven Apollo landings all occurred during the early lunar morning. This allowed the astronauts to spot craters and boulders more easily, without the harsh sunlight during the day.

PHYSICAL DATA

DISTANCE FROM EARTH TO THE SUN: = 149,597,000 km (light from the sun takes about 8 min 20 secs to reach the Earth)

DIAMETER: = 1,390,400km (109 Earths would fit across the sun's disc)

Once thought to be exceptional, our sun is now known to be an ordinary and rather nondescript star situated on the outskirts of an average galaxy. It is easy to understand how successive generations held the sun in particular esteem – it is the source of life on Earth; it warms up our world and gives light during our daytime.

Nevertheless, in the grand scheme of the universe, the sun is just another middle-aged, Main Sequence star. Its stellar classification, G2, denoting its yellowish coloring and its surface temperatures of 5000–6000 degrees Kelvin, is not extraordinary either; there are countless G2-type stars. Perhaps the sun's only distinction might prove to be that one of its planets harbors intelligent life.

FUSION

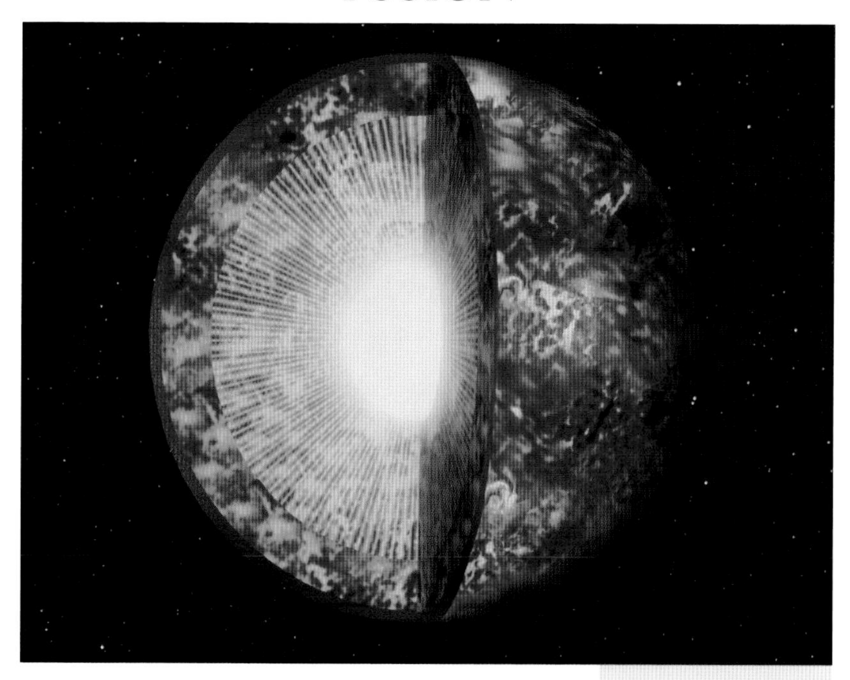

Initially it was thought that the sun was just one giant ball of fire, burning much like a candle. This theory is not tenable, because if the sun were burning like a candle, it would have been exhausted years ago. Another theory suggested that the sun's huge gravitational pull, the result of its large mass, was being converted into energy. Again, this does not account for the age of the sun, which must be more than 4.6 billion years old, the age of the oldest known rocks on the Earth.

At the beginning of the twentieth century, the German-born Jewish physicist Albert Einstein finally provided the answer when he stated that mass and energy were interchangeable, $E = mc^2$, where E is energy, m is mass and c is the speed of light. In the sun hydrogen nuclei collide with one another to create helium atoms. In this process, named fusion, mass is transferred into energy. Fortunately for our descendants, this process will continue for some millions of years to come, until the last of the hydrogen in the sun is used up.

ABOVE: Computer graphic of the sun, showing its internal layered structure. The sun is a massive nuclear fusion reactor. The core (white) has a temperature of at least 14 million degrees Celsius where hydrogen atoms fuse into helium to release heat and light energy. At the surface (red) is the photosphere, about 300 kilometers thick. Its constant output of energy means that the sun loses about 4 million tons of mass each second.

PHYSICAL DATA

CIRCUMFERENCE:
= 4,379,000km
(Earth = 40,075km)

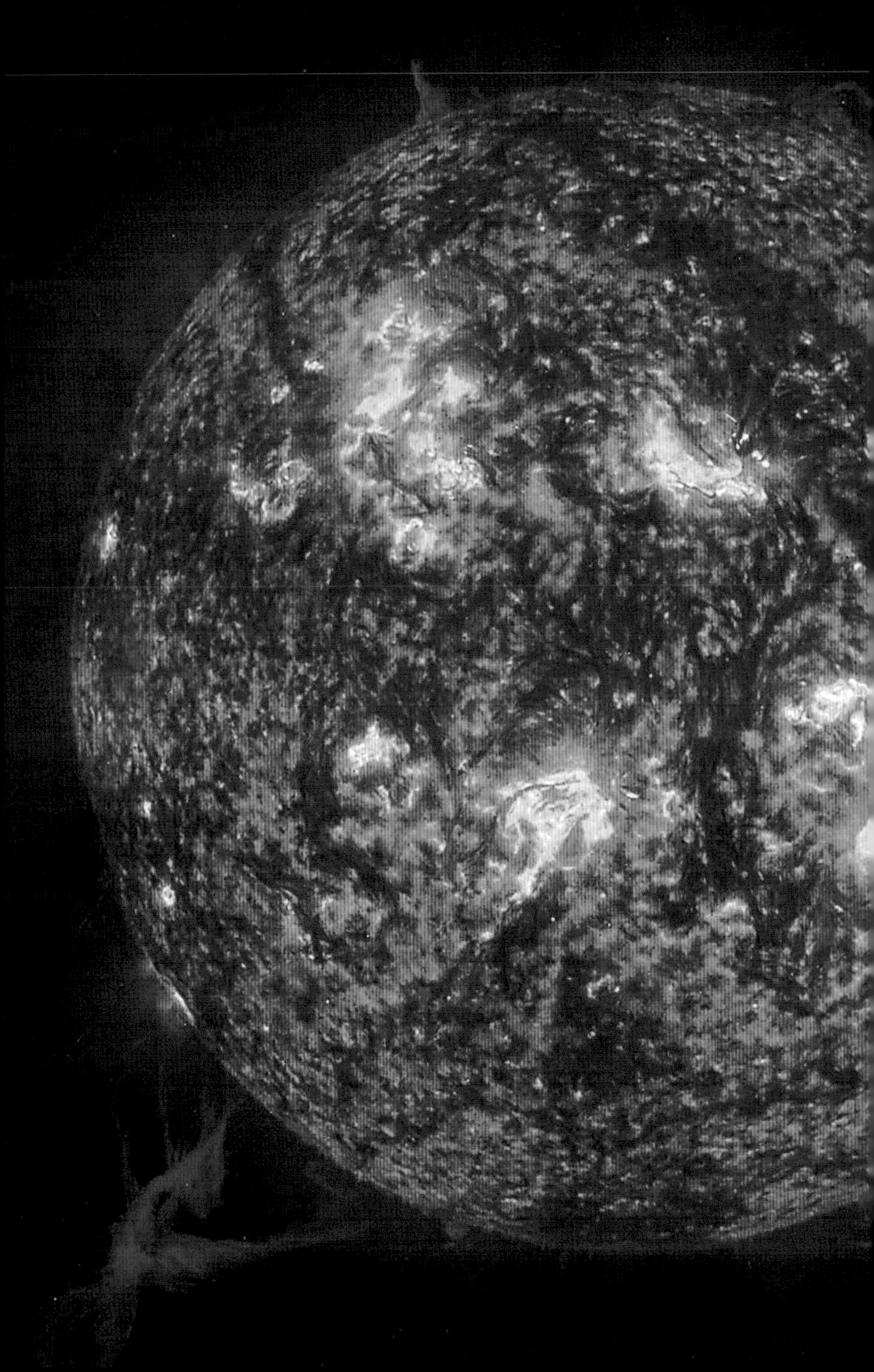

SOLAR LAYERS

The bright white disc that appears in the sky is the layer of the sun called the photosphere. The photosphere is essentially the surface of the sun – it emits the visible light that we see here on Earth. Temperatures in the photosphere of our sun are in excess of 5500 degrees Celsius. The surface of the sun has a granulated appearance, as hot gases rise to the surface and cooler ones fall back down to the convection zone below. Although these may appear as granules, relative to surface features on the Earth, they are vast in size. Although the photosphere is the outermost visible layer, there are others above it, often referred to as the solar atmosphere.

LEFT: Solar prominence. SOHO (Solar and Heliospheric Observatory) image of a huge twisted prominence (lower left) in the corona of the sun. The prominence is a massive cloud of plasma confined by powerful magnetic fields. If it breaks free of the sun's atmosphere, such an event can cause electrical blackouts and auroral storms if directed towards Earth.

TIMELINE

c400B.C. - Early people believed that the Earth was flat and that the Sun was a god. The Greek philosopher Anaxagoras realized that the Sun must be a large body, far from the Earth. He estimated the Sun's diameter at 56 kilometers. Anaxagoras' ideas disagreed with the religious beliefs of his time. His life was threatened, and finally he was exiled from Athens.

150 - The astronomer Ptolemy of Alexandria declared that the Earth was a stationary body in the center of the universe. He believed that the Sun, Moon, planets, and stars all circled the Earth.

1543 - Copernicus presented a planetary model with the Sun at the center of all planetary motions.

1610 - Galileo observes sunspots with his telescope.

SOLAR ATMOSPHERE

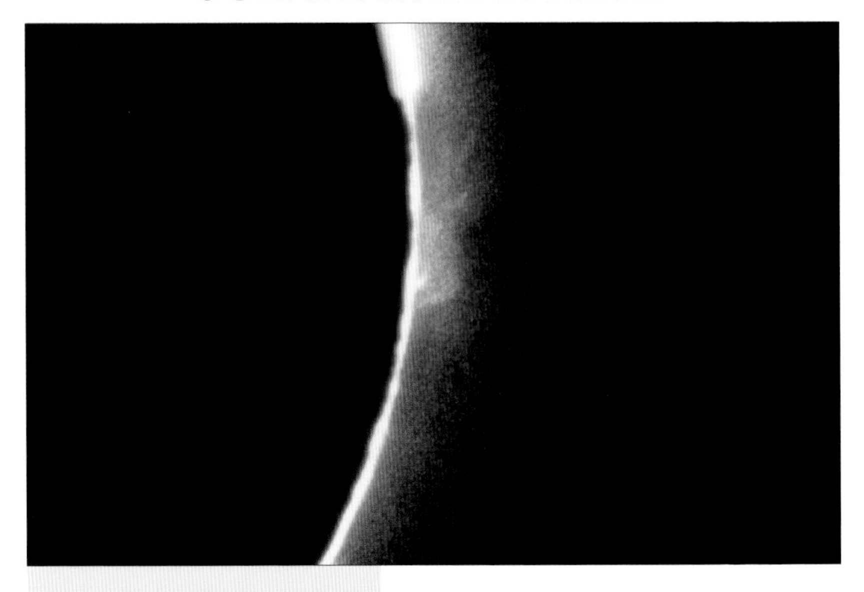

ABOVE: A total eclipse of the sun showing Bailey's Beads, prominences, and the solar chromosphere. The dark Moon is just covering the face of the sun. A narrow sickle of the brilliant light of the sun is seen through valleys and irregularities on the Moon's surface, forming what are known as Bailey's Beads. The sun's lower atmosphere, the chromosphere, is seen as a red arc. Flame-like red prominences are made up of incandescent gas and may exceed 500,000 kilometers in length.

PHYSICAL DATA

SURFACE AREA: = 6,087,799,000,000km^2 (11,990 times the surface area of the Earth)

ORBITAL PERIOD: The sun takes about 225 million years to make one revolution around the center of the Milky Way. During this period the sun travels about 10 million times as far as the distance between it and the Earth

ROTATIONAL PERIOD: 25 days 9 hours 7 mins

The solar atmosphere is divided into three layers, the chromosphere, the transitional zone, and the corona. All three are outshone by the photosphere so can only be observed as visible light during a solar eclipse. The chromosphere is a pinkish layer where much of the stormy weather on the sun is detected. Spicules, shortlived jets of gas, protrude outward from the interior of the sun through the chromosphere. The transition zone is where the temperature rises sharply from the 5500 degree temperatures in the photosphere and chromosphere to levels in excess of one million degrees in the corona. The word corona comes from the Latin for crown, and during an eclipse it becomes obvious where this analogy comes from. The corona looks like a great crown, extending in all directions millions of kilometers away from the disc. It is caused by sunlight scattering onto, and projecting off, electrons and interstellar dust.

SOLAR INTERIOR

Beneath the photosphere is the sun's interior. The sun's core undergoes such intense heat and pressure that the process of fusion is unremittingly generated. Surrounding the core is the radiation zone. Photons emitted from the core in the process of fusion collide with ions in this layer and transfer small amounts of energy. The radiation zone eventually gives way to the convection zone, where the temperature is lower – particles are heated by convection. Less dense hotter gases rise to the surface of the sun, where they cool and sink back down to the depths of the convection zone where they are heated and rise to the surface once again.

ABOVE: Annular eclipse of the sun. An annular eclipse of the sun is when the moon in the middle of the eclipse conceals the central part of the sun's disc, leaving a complete ring of light around the border.

MISSIONS TO THE SUN

1990 - MISSION NAME: ULYSSES

Ulysses was launched towards Jupiter from the Space Shuttle Discovery and used Jupiter's gravity to break out of the Ecliptic plane and fly over the Sun's polar regions.

1995 - MISSION NAME: SOHO

The international Solar and Heliospheric Observatory (SOHO) is a joint project of the European Space Agency (ESA) and NASA. It keeps the Sun under constant observation and has discovered dozens of comets. SOHO's data about solar activity is used to predict solar flares that could potentially damage satellites.

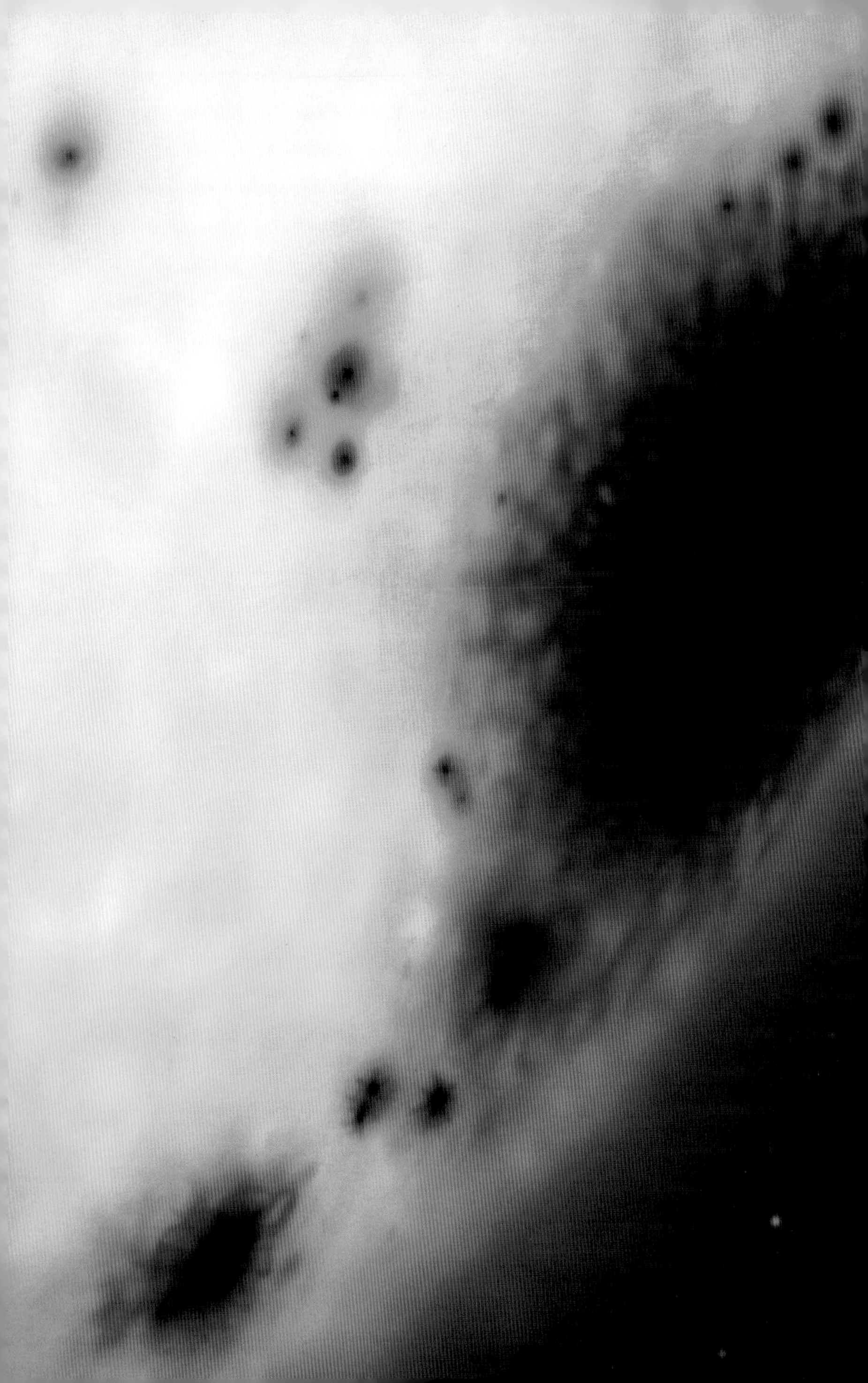

SUNSPOTS

In 1611 Galileo first reported seeing sunspots on the surface of the sun. These are minor regions, or spots, where the temperature of a small area of the photosphere is markedly cooler than its surroundings. This is a result of a strong, localized magnetic field that prohibits heated particles from rising up to the surface of the sun. When sunspots are looked at in more detail, it can be seen that while the center, the umbra, is a blackish color, the surrounding spot, the penumbra, is gray. A solar cycle occurs every eleven years, when the sun is at its most active. At this time, known as the solar maximum, more than one hundred sunspots can be seen. However, during the solar minimum, there can often be no sunspots at all.

Never look directly at the sun, even when it seems that the sun's brightness has been reduced by cloud, sunglasses or some other method. Certainly do not look at the sun through binoculars, telescopes or cameras. Galileo went blind in the later years of his life, possibly as a result of his solar observations. The best way of observing the sun is by projecting its image onto a screen through the lenses of binoculars or a telescope.

LEFT: Computer illustration of several large, dark sunspots. Sunspots are areas on the sun's surface which are around 2000 degrees Celsius cooler than their surroundings, causing them to appear darker. The spots here have a dark region, the umbra, surrounded by a less-dark region, the penumbra. Sunspots are a sign of an active sun.

TIMELINE

c800B.C. - The first plausible recorded sunspot observation in China.

1610 - Galileo observes sunspots with his telescope.

1650-1715 - Maunder Sunspot Minimum; a period when there was a dearth of sunspot sightings.

1908 - First measurement of sunspot magnetic fields taken by American astronomer George Ellery Hale.

1954 - Galactic cosmic rays found to change in intensity with the 11-year sunspot cycle.

SOLAR FLARES

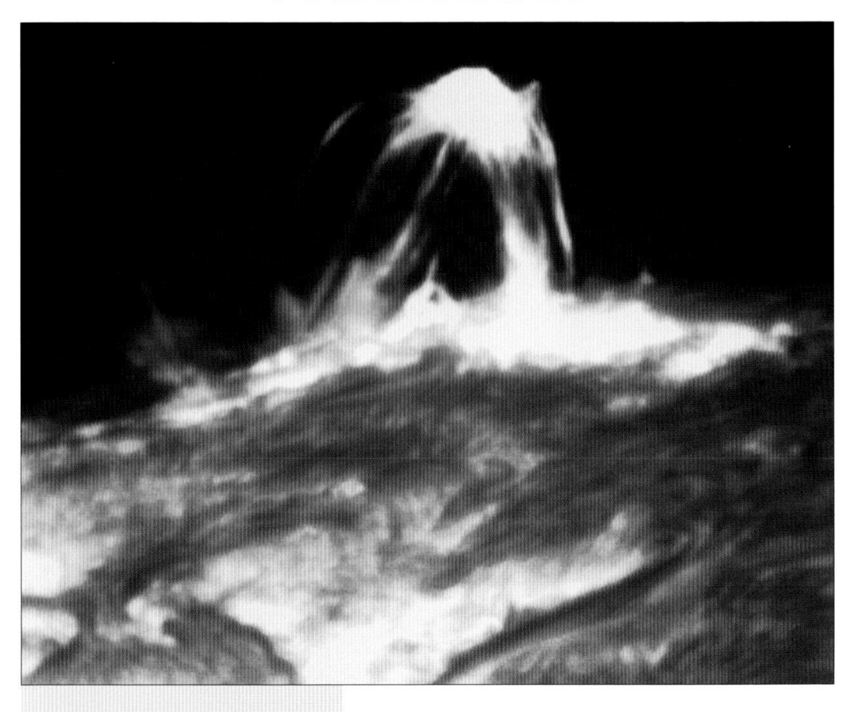

ABOVE: Colored image of a solar flare. The flare is erupting from the chromosphere, or photosphere. Solar flares are explosive eruptions associated with active regions of the sun. The temperature within a flare, which normally lasts for a few minutes, can reach hundreds of millions of degrees Celsius.

TIMELINE

1859 - First observation of a solar flare by amateur astronomer Richard Carrington.

1956 - Largest observed solar flare occurred.

1982 - First observations of neutrons from a solar flare by Solar Maximum Mission (SMM).

Solar flares were first detected by a British astronomer, Richard Carrington, during a particularly large flare in 1859. A solar flare occurs when energy that has built up in the sun's interior is suddenly released. The abrupt discharge of such an incredible amount of energy manifests itself in all forms across the electromagnetic spectrum as radio waves, gamma rays and X-rays, although it is difficult to see a flare on the visible part of the spectrum because the glare from the photosphere is too great. If particles from a solar flare come into contact with the Earth's magnetic field they can disrupt communications such as radio and satellite links. Additionally, these particles are potentially lethal to astronauts working outside the Earth's atmosphere.

SOLAR WIND

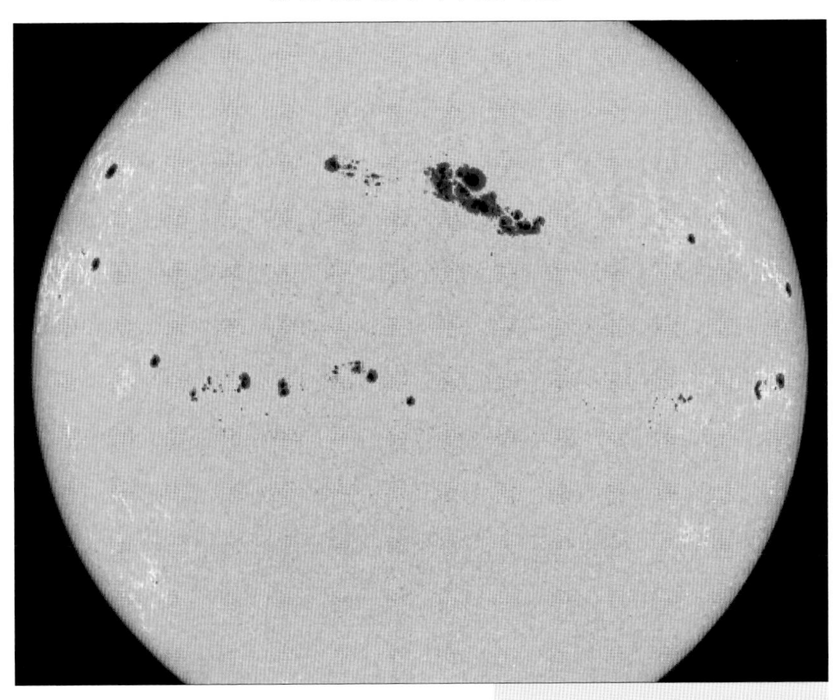

Some particles are heated to such high temperatures that they are able to escape the sun's gravity. These move at rapid speeds of around 300 miles per second as a stream of particles called solar wind. Many of the planets in the solar system generate a magnetic field, which protects them from the solar wind, preventing it reaching the planet. Sometimes, solar particles are caught in a planet's magnetic field and drawn to the planet's poles, where they interact with atmospheric gases causing fantastic light displays named auroras, like Earth's Aurorae Borealis and Australis. Auroras increase in frequency when the sun is at its period of maximum activity, which occurs every eleven years.

ABOVE: An enormous sunspot group on the surface of the sun. Sunspots are relatively cool areas on the sun's surface associated with strong magnetic fields. At the end of March 2001, the area of this sunspot group was over 13 times that of the surface area of the Earth, making it the largest sunspot group seen during that solar cycle.

PHYSICAL DATA

MASS: = 1.99 x 10^{30}kg (Approx 333,400 more massive than the Earth and contains 99.86% of the mass of our entire Solar system)

VOLUME: = 1.14 x 10^{18}km^3 (equivalent to 1,300,000 Earths)

DENSITY: = 1.4 g/cm^3 (about one quarter of the density of the Earth)

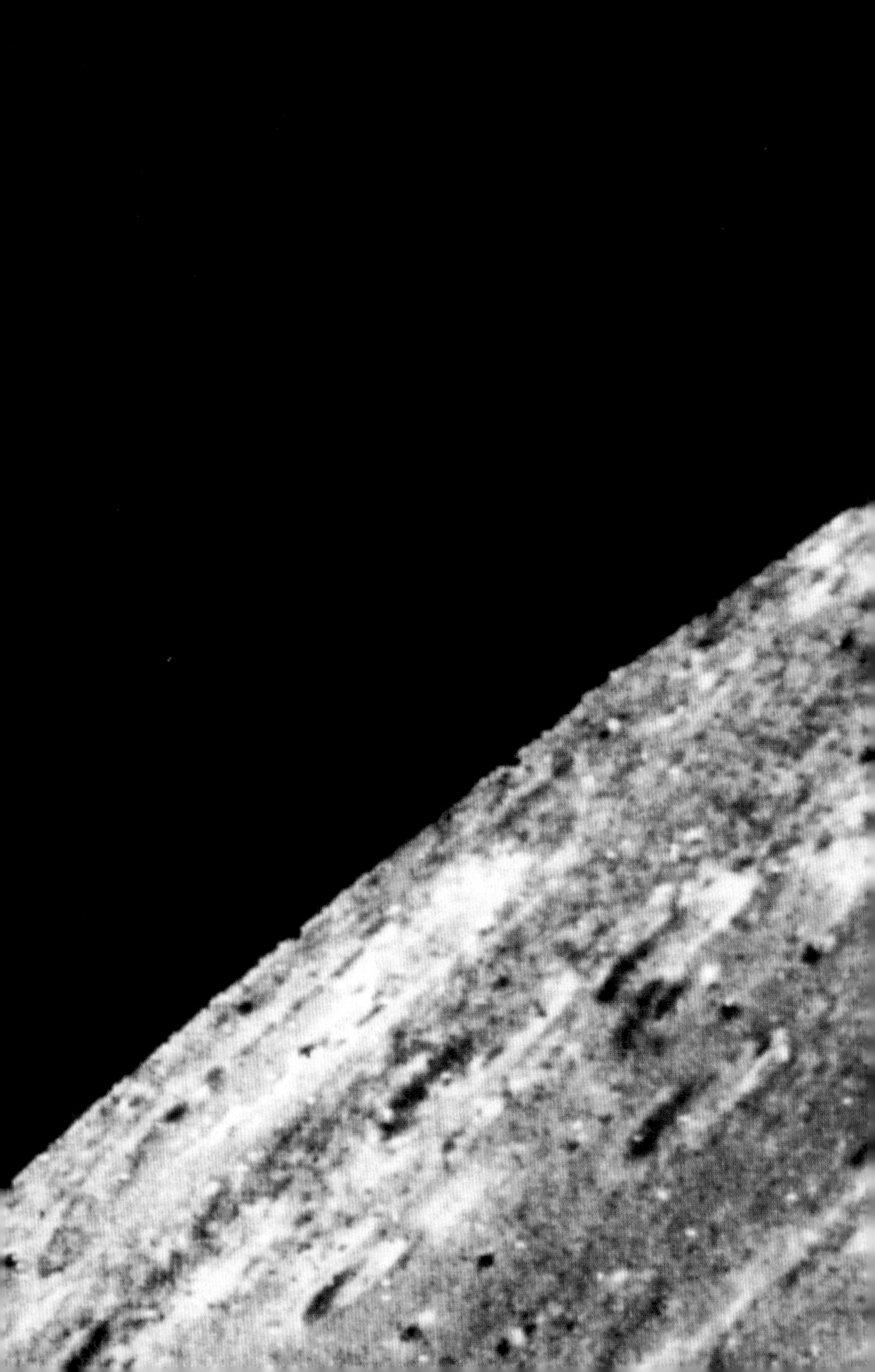

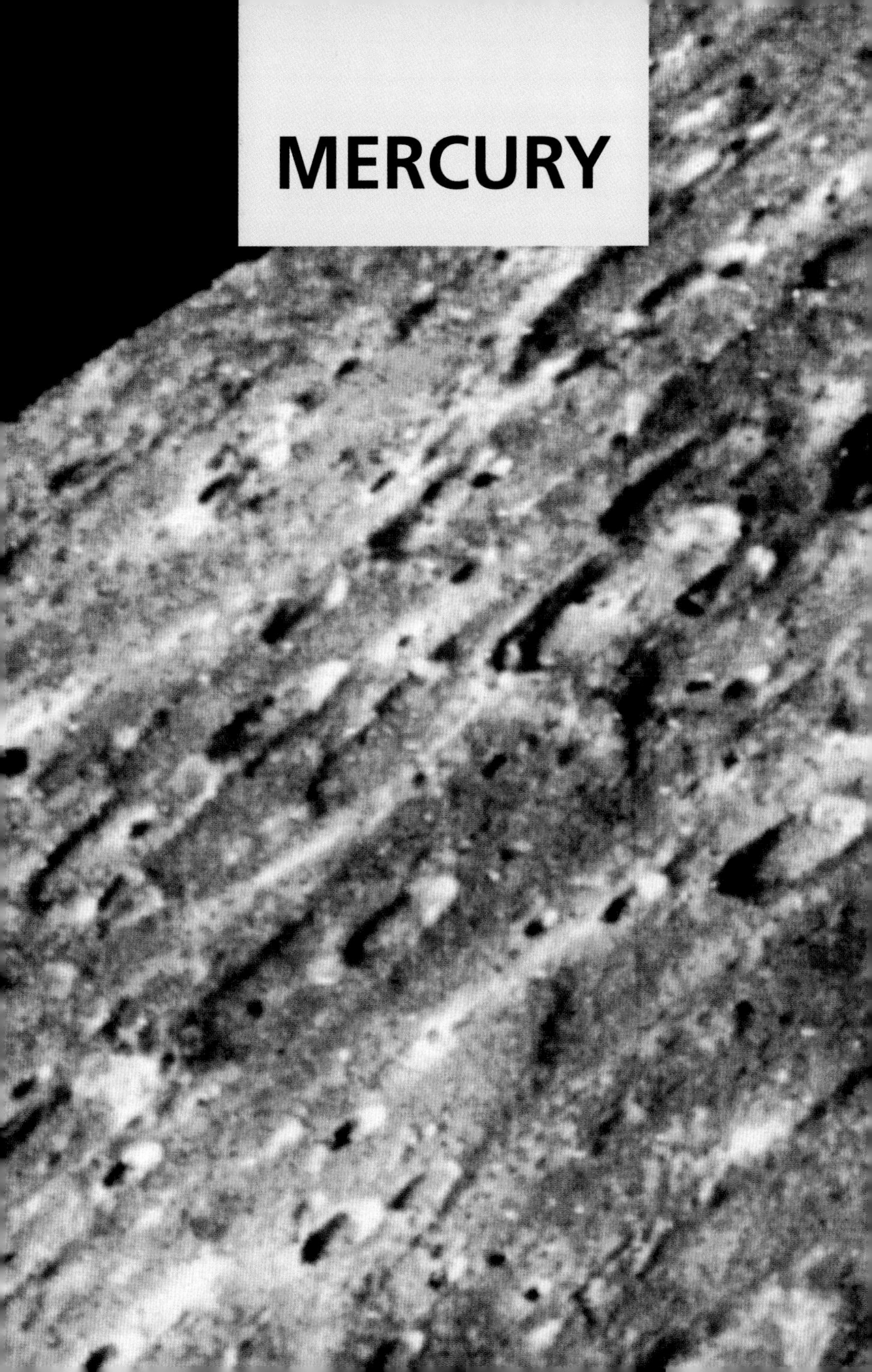
MERCURY

THE FASTEST PLANET

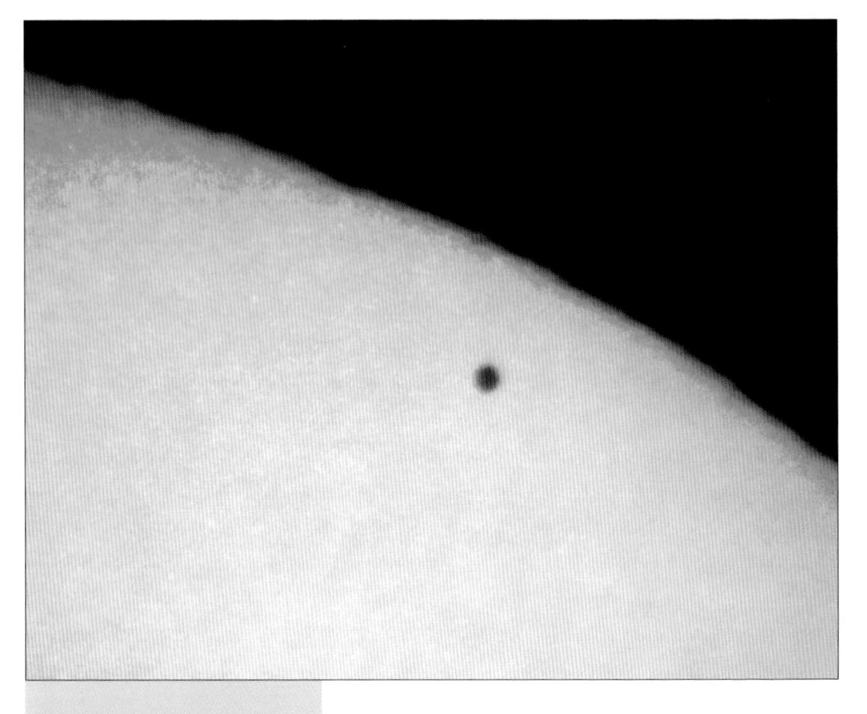

ABOVE: Optical photo of the planet Mercury transiting the sun. Mercury appears as a small black dot silhouetted against the brillant disc of the sun. Thirteen times a century the planet Mercury passes directly between the Earth and the sun.

TIMELINE

c385B.C. - Greek astronomer Heracleides became the first person to suggest that Mercury orbits the Sun.

1610 - Italian astronomer Galileo made the first telescopic observation of Mercury.

1631 - French astronomer Pierre Gassendi made the first telescopic observations of the transit of Mercury across the face of the Sun.

Very little was known about Mercury, the nearest planet to the sun, until Mariner 10 visited in 1974. Beforehand, it had been very difficult to observe through a telescope due to its small size and proximity to the sun, so difficult in fact that it is said Copernicus never managed to see Mercury for himself. Mercury's closeness to the sun means that it is best observed during the hours of dawn and dusk, because during such twilight hours it is not overpowered by the luminosity of the sun. To view Mercury, it is best to wait until it is at maximum elongation – the greatest angular distance between an inferior planet and the sun, as seen from Earth. Mercury's maximum elongation is a mere 270, meaning even when it appears to be at its furthest distance from the sun, a would-be observer must always seek to find the planet in the direction of the sun.

MESSENGER TO THE GODS

The planet was named by the Romans after Mercury, the messenger to the gods, because they likened its quick movement across the sky to the speed at which the messenger was believed to move. As the closest planet to the sun it completes one orbit in a mere 88 days, at breakneck speeds of 50km per second, making it the fastest planet in the solar system. It was originally thought that the time it took Mercury to rotate on its axis was also 88 days, meaning it would only ever present one face to the Sun. However, the far side of Mercury was deemed too warm to have endured perpetual darkness and instead the planet's rotation period has been recalculated to be 59 Earth-days.

ABOVE: Mariner 10 photograph of Mercury, the innermost planet, showing its heavily cratered surface. Mariner 10 was the first spaceprobe to visit the planet, with three separate encounters during 1974-5.

PHYSICAL DATA

DISTANCE FROM SUN: = 57,909,175 km (Earth to the sun = 149,597,000 km)

ORBITAL PERIOD: = 88 days

ROTATIONAL PERIOD: = 59 Earth-days

Mercury completes three rotations for every two orbits round the sun. If you wanted to experience a Mercury day you would have to stay up for 136 Earth-days

THE FIRST PROBE

ABOVE: Composite of images of Mercury seen from the Mariner 10 spacecraft. Numerous meteorite impact craters are seen.

PHYSICAL DATA

DIAMETER: = 4,878km (approx 40% of Earth)

CIRCUMFERENCE: = 15,329km (Earth = 40,075km)

SURFACE AREA: = 74,800,000km^2 (Earth = 510,072,000km^2)

Mariner 10 was launched in November 1973. Previously, it had been considered too costly to build rockets powerful enough to break Earth's orbit and lock into Mercury's even faster orbit, but a new method was used: the gravitational pull of Venus was employed to assist Mariner 10 on its mission to Mercury, the so-called gravity-assist or "slingshot" maneuver. Mariner 10 flew past Mercury twice in March and September 1974, and then once again in March 1975. Together the three flybys managed to cover just under half of the surface of the planet. Mariner 10 unexpectedly identified a magnetic field. It was thought that the planet's core would have solidified long ago, but the existence of a magnetic field suggests that Mercury has a large metallic core that remains partially molten.

THE FUTURE PROBES

Given that Mariner 10 performed only three fleeting flybys, there is much that remains a mystery about Mercury. A new mission to Mercury, the aptly named MESSENGER mission was launched in August 2004. MESSENGER stands for MErcury Surface, Space ENvironment, GEochemistry, and Ranging. It is destined to rendezvous with Mercury in March 2011. The European Space Agency also intends to send a probe, BepiColombo, to Mercury early in the next decade. The mission will have three parts: a planetary orbiter, a magnetospheric orbiter and, most excitingly, a lander, fitted with a camera, which will beam back the first photographs from the planet's surface.

ABOVE: Mosaic photograph showing the heavily cratered surface of Mercury.

PHYSICAL DATA

MASS: = 3.30×10^{23}kg (If Earth were the size of a baseball, Mercury would be the size of a golfball)

VOLUME: = 6.1×10^{10}km^3(approx 5% of Earth)

DENSITY: = 5.4 g/cm^3 (Mercury is the densest planet in the solar system after the Earth)

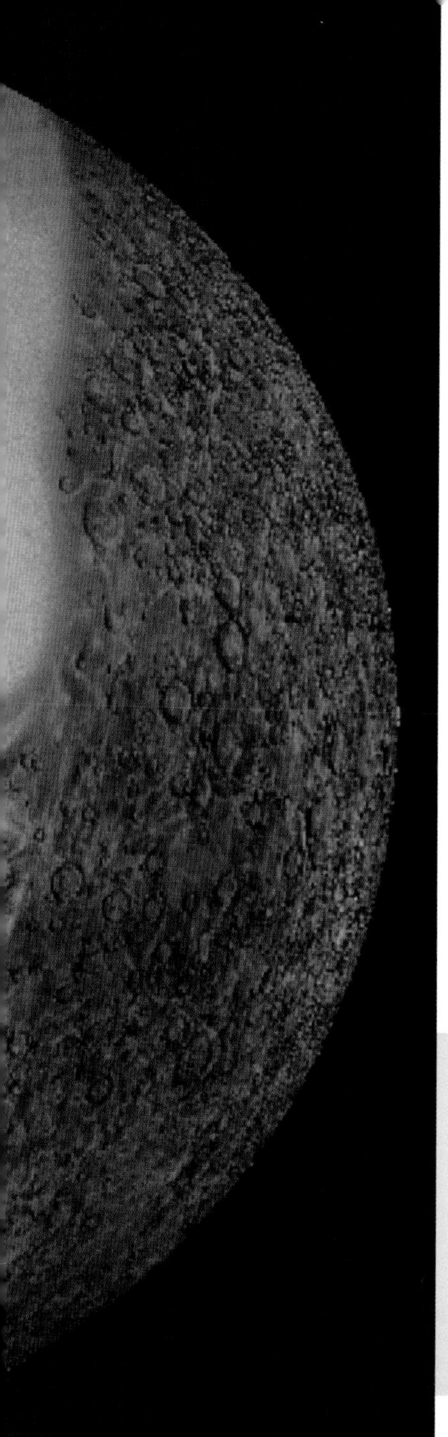

THE SURFACE

When Mariner 10 arrived, it found that Mercury was a cratered world, much like the Earth's moon. A massive crater was quickly identified and later named Kuiper. The planet's craters are most likely the result of meteorites colliding with the surface; one particularly large instance of a collision is the Caloris Basin. The basin is estimated to be 1300km across and the impact was so severe that mountains were thrown up on the opposite side of the planet.

The existence of smooth plains indicated that molten lava was once present, but the abundance of craters suggested that the planet was no longer geologically active, otherwise Mercury's pockmarked face would have been smoothed over. Massive cliffs, as high as three kilometers in places, are further indication that Mercury has cooled down. These cliffs are the likely result of the crust contracting as the planet cooled.

LEFT: Mariner 10 spacecraft mosaic image of Mercury. Areas for which data is missing are blank. The surface of Mercury is heavily cratered due to impacts from meteorites. It also has lines of cliffs which are up to 3km high and 500km long.

ATMOSPHERE

ABOVE: Internal structure of Mercury, cutaway artwork. The inset box shows Earth and Mercury at the same scale. Mercury is a rocky planet. Its interior comprises a large iron core. This underlies a thick silicate crust. Mercury has a very thin atmosphere, which does not trap much heat.

OPPOSITE: Mercury temperature map. Colored radio image of the surface of the planet Mercury, showing the range of temperatures across its surface. The hotter regions are represented by the yellow color. Mercury has two hot poles on its equator, produced by the intense heat of the nearby sun.

Mercury's atmosphere is so tenuous that scientists often term it an exosphere instead of an atmosphere. The exosphere is the outermost layer of the Earth's atmosphere where molecules have enough velocity to escape the planet's gravitational pull. As the whole of Mercury's atmosphere is like the outermost layer of Earth's atmosphere, most of its atoms and molecules are lost into space. When Mariner 10 visited Mercury, helium, hydrogen and oxygen were discovered in the atmosphere. The following decade, in the 1980s, potassium and sodium were also discovered. Mercury's magnetosphere is able to capture ions from solar winds, ensuring the atmosphere is constantly replenished, in spite of the rapid loss of molecules into space.

TEMPERATURE

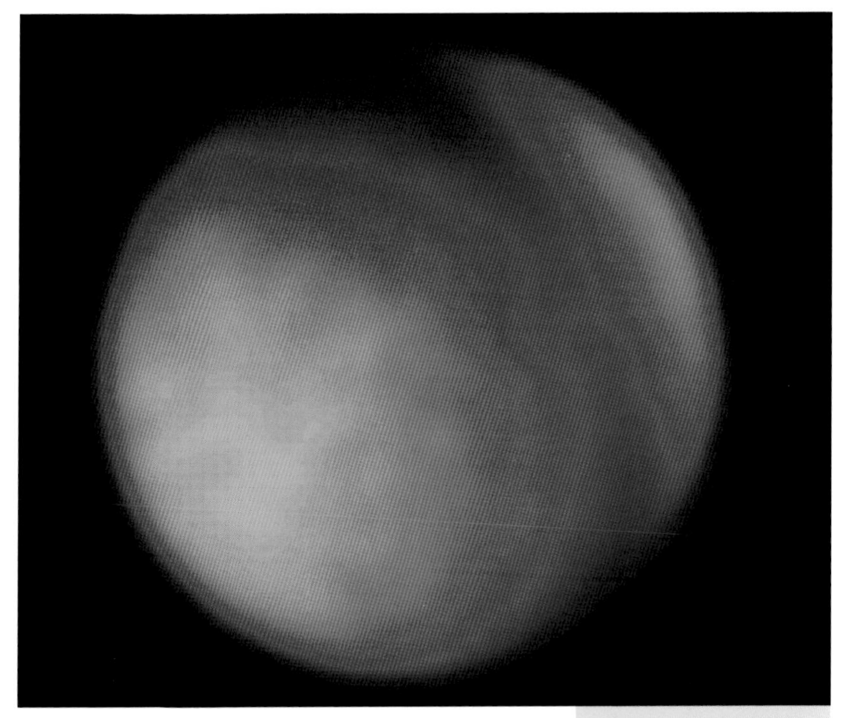

Mercury has a very weak atmosphere, with the resulting effect that the planet retains very little light; therefore, the skies would be dark in spite of daytime temperatures reaching as high as 350 degrees Celsius. During the long night, temperatures can drop as low as minus 170 degrees Celsius. As Mercury spins on an axis with almost no tilt whatsoever, seasons do not exist, as the sun remains strongest at the equator all year long. With such high daytime temperatures the existence of water in any form is highly unlikely. Nevertheless, in 1991, scientists bounced radio waves off the surface of Mercury, in an experiment which produced an anomalous reading around the north pole, perhaps indicating that frozen water might be present, hidden deep inside craters, safe from the sun's rays because Mercury's perpendicular axis means that the sun cannot penetrate inside.

PHYSICAL DATA

SURFACE GRAVITY: = 3.7m/s^2 (Mercury's smaller mass makes its force of gravity only about a third as strong as that of Earth. An object that weighs 100kg on Earth would weigh about 38kg on Mercury)

ATMOSPHERE: Potassium, sodium, oxygen, argon, helium, nitrogen, hydrogen

ATMOSPHERIC PRESSURE: trace (1.03kg/cm^3 on Earth)

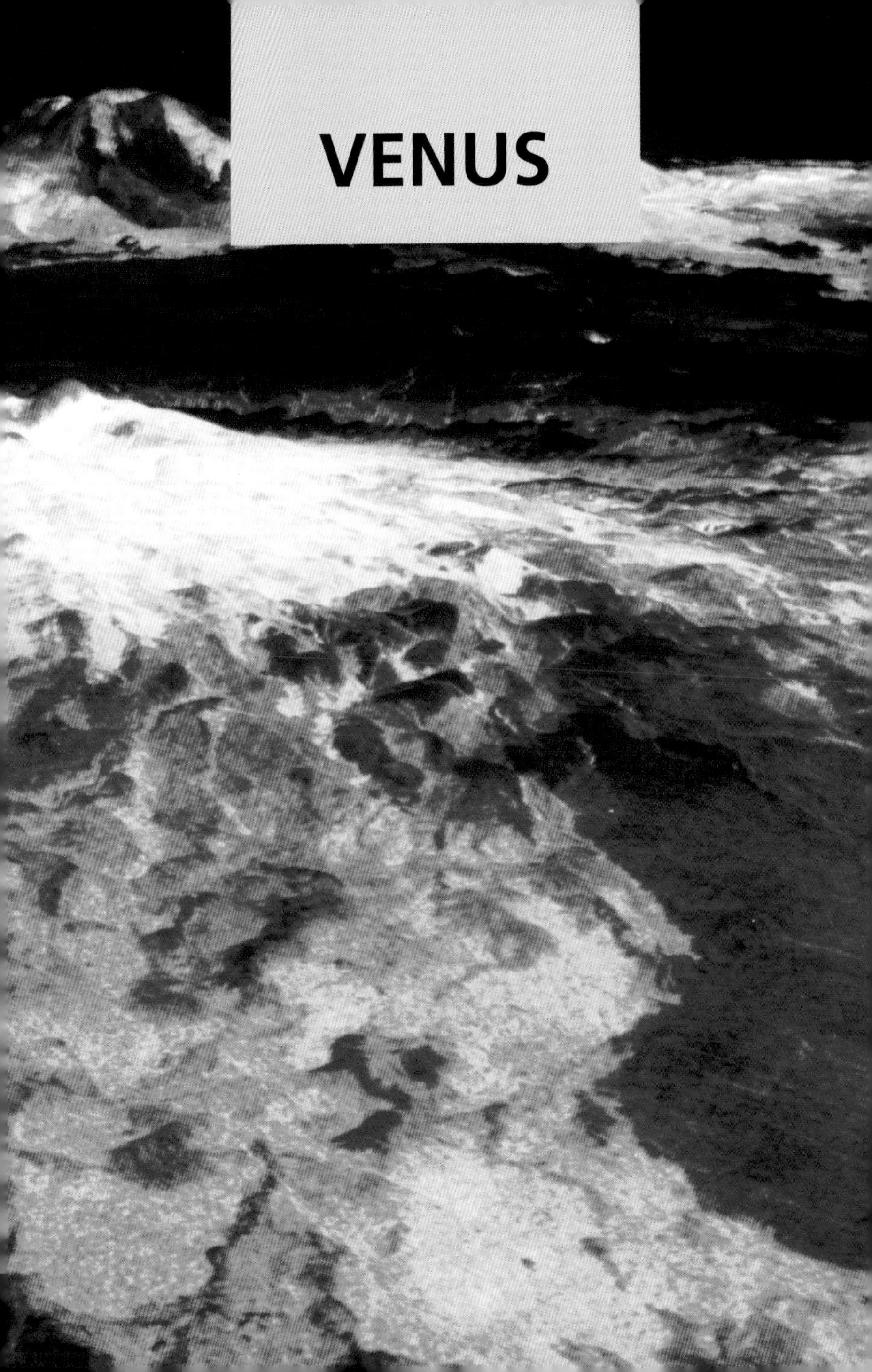

VENUS

THE BRIGHTEST PLANET

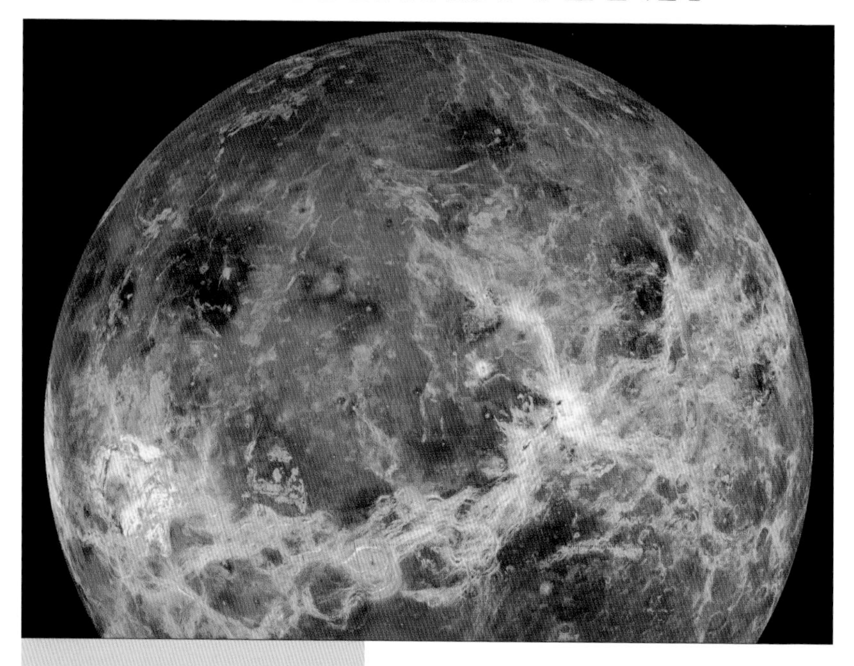

ABOVE: False-color projection of the surface of the western hemisphere of Venus. The north pole is at the top of the image. The bright band across the disc shows the extent of Aphrodite Terra, a "continent" about the size of Africa. At the left edge of the disc is the largest upland region. Just below this is a circular valley-like feature called Artemis Chasma.

TIMELINE

1610 - Galileo observed that Venus progresses through phases similar to those of the Moon.

1666 - French astronomer Cassini made the first measurements about the rate at which Venus spins on its axis.

1911 - US astronomer Vesto M. Slipher used spectral analysis to show that the rate at which Venus rotates is longer than one day.

The Romans named the second planet in the solar system Venus, after their goddess of love and beauty, perhaps because it shines gloriously in the evening or morning sky. Its dense cloud cover is a great reflector of light, which makes Venus the brightest planet in the solar system. It is so bright that Venus can sometimes be seen during Earth's daylight hours – the only celestial body, aside from the moon and the sun, to be visible during an Earth-day. Venus is closer to the sun than Earth and, as such, when searching for it in the sky, it must be looked for in the direction of the sun. Venus's orbit does not hug the Sun as tightly as Mercury's meaning it is not as easily outshone by the Sun and appears so brightly in the night sky that it has historically been given the titles of both Morning Star and Evening Star.

THE SISTER PLANET

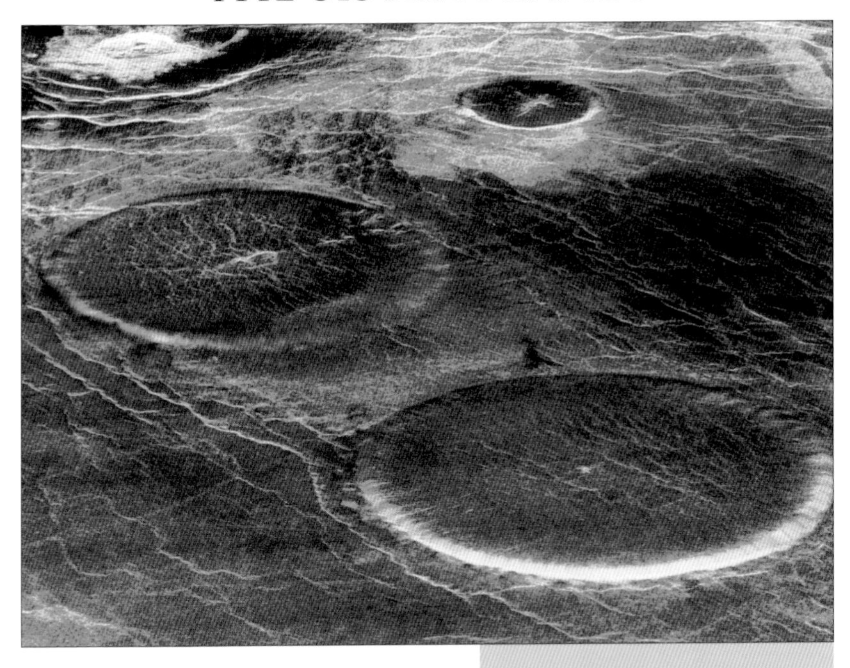

Venus is often referred to as Earth's sister planet, not only because they are one another's nearest neighbors in the solar system, but also because they are very similar in size – Venus and Earth have similar diameters, surface areas, volumes and masses. But that is where the similarities end. The average surface temperature is between 400 and 500 degrees Celsius. Such temperatures, which could melt lead, could almost certainly not sustain plant or animal life. Soviet and NASA probes charted the Venusian atmosphere and found that more than 90% was made up of carbon dioxide. Today that figure has been refined to an estimated 96% of Venus's atmosphere comprising the gas.

ABOVE: Artwork of volcanic domes on Venus. Venus has the hottest planetary surface in the solar system, with temperatures of nearly 500 degrees Celsius. This is due to its carbon dioxide atmosphere that traps the sun's heat. Layers of sulfuric acid clouds cover the planet.

PREVIOUS PAGES: Computer-generated false-color 3D view of a Venusian landscape, looking towards Sapas Mons.

PHYSICAL DATA

MASS: = 4.87×10^{24}kg (about 80% of mass of Earth)

VOLUME: = 9.3×10^{11}km^3

DENSITY: = 5.2 g/cm^3 (A "portion" of Venus would weigh a little less than an equal-sized "portion" of Earth)

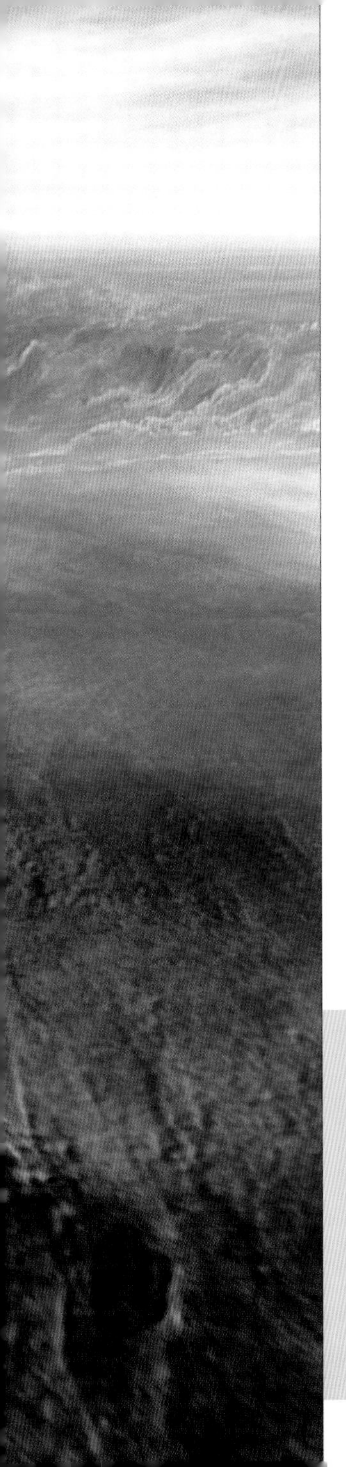

THE GREENHOUSE EFFECT

Venus's high temperatures cannot be explained by being closer to the sun than the Earth because Mercury's highest recorded temperatures fall almost 100 degrees Celsius short of the average Venusian temperature. The explanation is that Venus is being heated like a greenhouse. All planets with atmospheres have a naturally occurring greenhouse effect. Energy from the sun is able to penetrate through the gases in the atmosphere on its way in to the planet because such solar energy has very short wavelengths. Waves that reach the surface are either absorbed or reflected. Those reflected are emitted as infrared, which has a longer wavelength than the incoming rays, meaning their escape from the planet's atmosphere is not as easy as was their entrance. While some infrared rays manage to escape into space, many are absorbed by atmospheric gases which reflect the infrared back towards the surface, thus heating up that planet. It is the sheer abundance of carbon dioxide in Venus's atmosphere, and not its thick cloud cover, that traps the heat, leading to a greenhouse effect that is more exaggerated than on any other planet in the solar system.

LEFT: Computer-generated perspective view, looking east towards the Maxwell Montes range on Venus. Maxwell Montes (just above center) is the tallest mountain range on Venus, reaching 11km above the mean planetary radius.

PHYSICAL DATA

SURFACE GRAVITY: = $8.9m/s^2$ (An object that weighs 100kg on Earth would weigh about 91kg on Venus)

ATMOSPHERE: Carbon dioxide (96%), nitrogen, argon, carbon monoxide, neon, sulfur dioxide

ATMOSPHERIC PRESSURE: $93kg/cm^3$ ($1.03kg/cm^3$ on Earth: the pressure is so intense on Venus that standing on the surface would feel like the pressure felt 900 meters deep in Earth's oceans)

TRANSIT OF VENUS

ABOVE: Transit of Venus. Venus is the black dot seen on the sun at lower left. A transit occurs when Venus passes in front of the sun as seen from Earth. This was the first transit since 1882. As viewed from Earth, only Mercury and Venus can transit the sun.

PHYSICAL DATA

DISTANCE (AVERAGE) FROM SUN:
= 108,208,930 km
(Earth to the sun = 149,597,000 km)

AVERAGE SPEED ORBITING THE SUN:
= 35km/sec (Earth = 30km/sec)

When Venus is at inferior conjunction it usually passes above or below the sun. However, rarely, when both Earth's and Venus's orbital planes cross, a rare sight, the transit of Venus across the sun, can be observed. The same occurs as with a solar eclipse of the Earth by the moon, but Venus is much farther away than the moon and the effect is far less pronounced. Instead, the small outline of Venus can be seen crossing the sun. A transit of Venus is a rare sight indeed. Transits occur in pairs, with over a hundred years difference between each couplet. The most recent transit was in 2004; before that, the last transit pair was in 1874 and 1882. The partner to the 2004 transit is expected in 2012 and will be best observed from the Pacific Ocean.

ROTATION

Venus is unique in that a Venusian day lasts longer than a Venusian year. It takes Venus 243 Earth-days to rotate on its axis, while it takes just 225 Earth-days for Venus to orbit the sun. Venus rotates on its axis from east to west, instead of the conventional direction from west to east. Therefore, on Venus the sun rises in the west and sets in the east, the reverse of what we are used to on Earth. It would appear that Venus is rotating in a retrograde fashion at an axial inclination of just two degrees. However, many astronomers believe that Venus is upside down, with its north pole 178 degrees from the perpendicular, which would mean that Venus is rotating in a direct, conventional fashion, but upside down.

ABOVE: Colored radar map of the eastern hemisphere of Venus. The map was created using data gathered by the Magellan radar-mapping spacecraft in 1990-94. The colors indicate the elevation of the surface relative to the mean planetary radius (6041km). Blue colors denote land below the mean radius; land above the mean radius is colored green.

PHYSICAL DATA

ORBITAL PERIOD: = 225 Earth-days

ROTATIONAL PERIOD: = 243 Earth-days

CLOUD COVER

ABOVE: False-color radar map of the western hemisphere of Venus. Four major radar-bright upland features are seen. Running down the image just right of center are Beta Regio (top), Phoebe Regio and the less-distinct Themis Regio (bottom). The bright area at center left is the eastern extent of the continent Aphrodite Terra.

OPPOSITE: Clouds over Venus. The general structure of these clouds has remained the same over many years. There are two dense bands of cloud near each pole. These high clouds are mainly sulfuric acid droplets.

With such high surface temperatures, the layer of clouds which obscure the Venusian surface form much higher up in the atmosphere than clouds do on Earth. Most clouds on Venus form in a belt between 45km and 65km above the surface, while clouds on Earth seldom reach much higher than 10km. The clouds comprise mainly sulfuric compounds which are moved rapidly round the planet by fast winds of up to 300 km/h. Such rapid winds are only to be found at the cloud tops; at the surface, winds travel at relatively modest speeds, just a few kilometers per hour. Venus also experiences rain, but unlike Earth's water-based rain, Venusian rain consists of sulfuric acid.

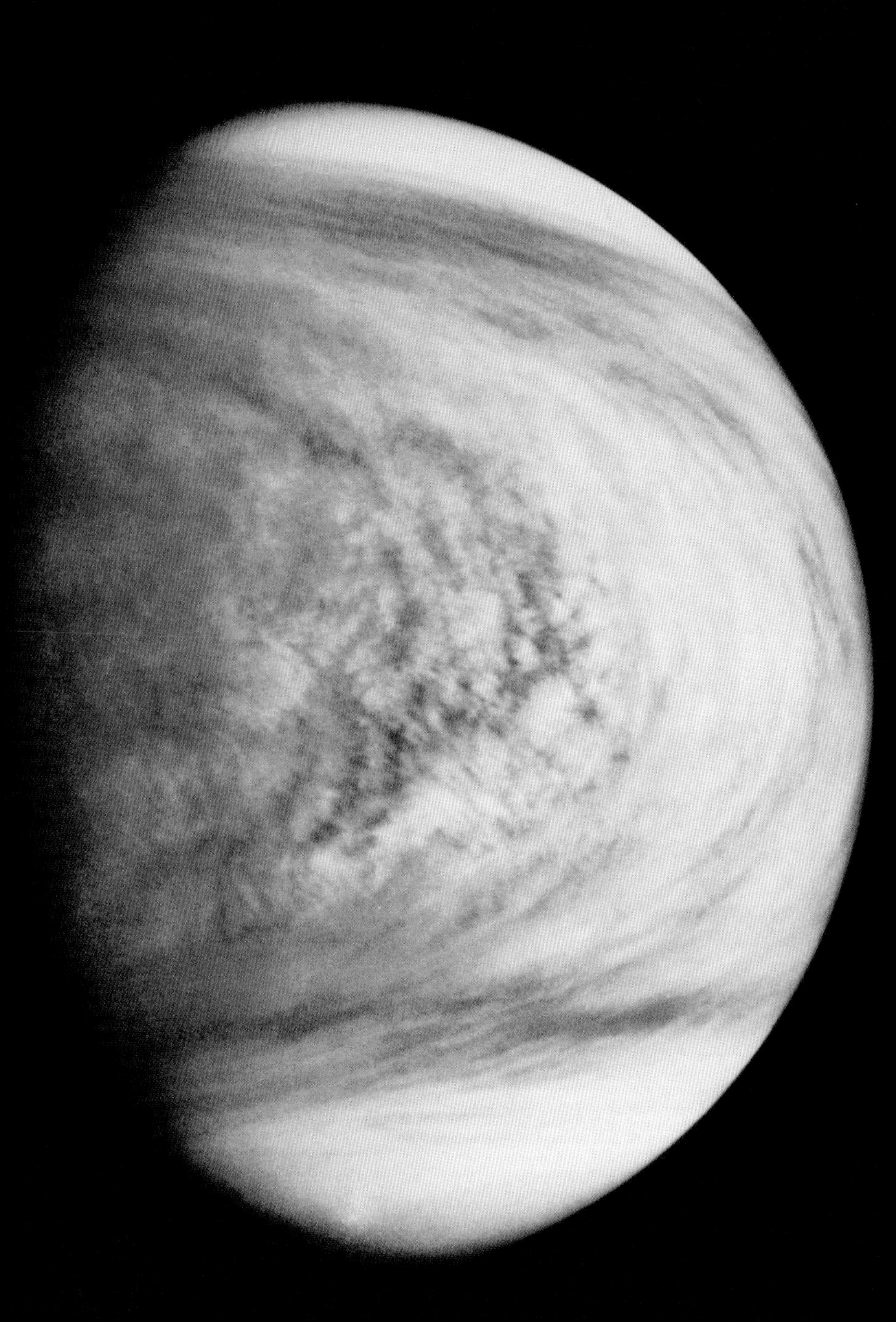

FINDING OUT-FIRST PROBES

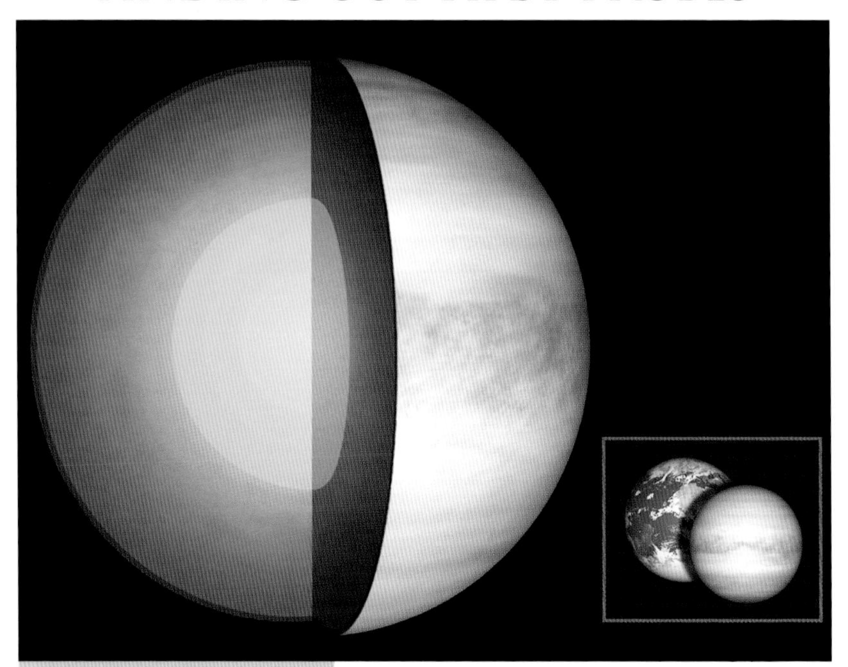

ABOVE: Cutaway artwork shows the internal structure of Venus. The inset at lower right illustrates Earth and Venus on the same scale, showing that they are around the same size. At the center of Venus is a primarily solid iron core, which underlies a thick mantle made mainly of silicate minerals.

PHYSICAL DATA

DISTANCE FROM THE EARTH:
= 38,150,900km

CRUST: = 30km thick (Earth = 70km)

AVERAGE TEMPERATURE (SURFACE):
= 400-500 degrees C
(Earth temperature ranges from -90 to 60 degrees C)

Prohibited by dense cloud cover from conducting clear observation of the Venusian surface, both the United States and the Soviet Union made it their aim to understand more about what lay beneath. The first successful interplanetary probe was Mariner 2 which reached Venus in 1962; while it mapped the atmosphere it did not penetrate the cloud cover. In 1970, the Soviet probe Venera 7 became the first successful landing of a spacecraft on a planet. Venera 7 relayed less than half-an-hour of data under the strenuously demanding conditions – temperatures of over 470 Celsius, and an atmospheric pressure assessed at ninety times that experienced on the Earth's surface. Humans would be crushed under the pressure immediately; this has ruled out a manned mission to Venus for the foreseeable future.

THE SURFACE

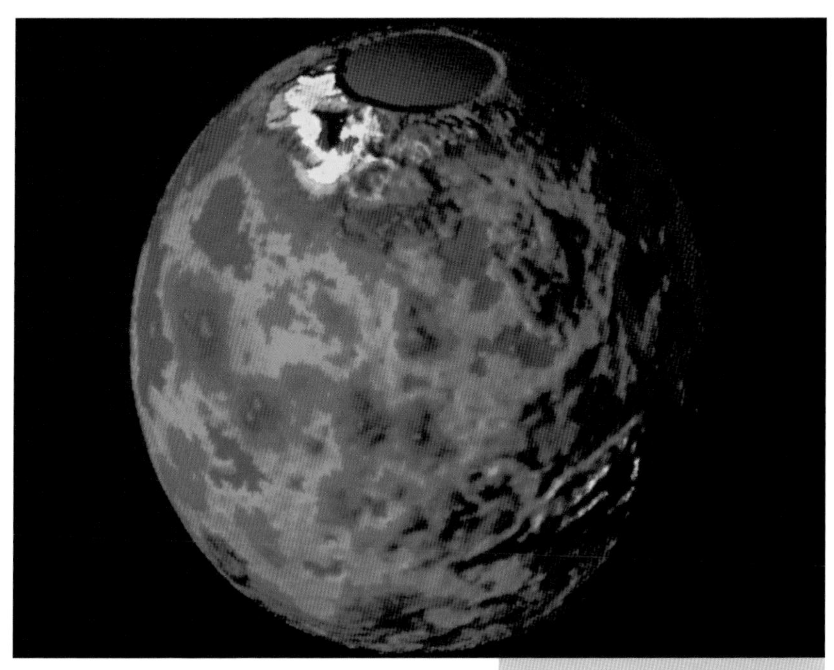

In 1975 two Soviet probes, Venera 9 and Venera 10 landed on Venus within one week of one another – both probes photographed the Venusian surface for the first time. The black and white pictures revealed rocky remnants of lava from volcanoes. A NASA probe, Pioneer Venus was dispatched to begin mapping the planet and identified two areas of highlands, the size of continents, later called Aphrodite Terra and Ishtar Terra. The highest point on Venus, Maxwell Montes, was also discovered. It is difficult to calculate its exact height, but when measured from the average surface elevation, it is thought to be around 11.5km tall, over 2km higher than Mount Everest.

ABOVE: Color-coded topographic map of the surface of Venus. The smooth circle at top marks the north pole. Most of the planet's surface is covered by relatively smooth plains, colored blue in this image. Two large highland regions are visible, Ishtar Terra (upper left) and Aphrodite Terra (lower right).

PHYSICAL DATA

DIAMETER: = 12,104km
(Earth = 12,756km)

CIRCUMFERENCE: = 38,025km
(Earth = 40,075km)

SURFACE AREA: = 460,200,000km^2
(Earth = 510,072,000km^2)

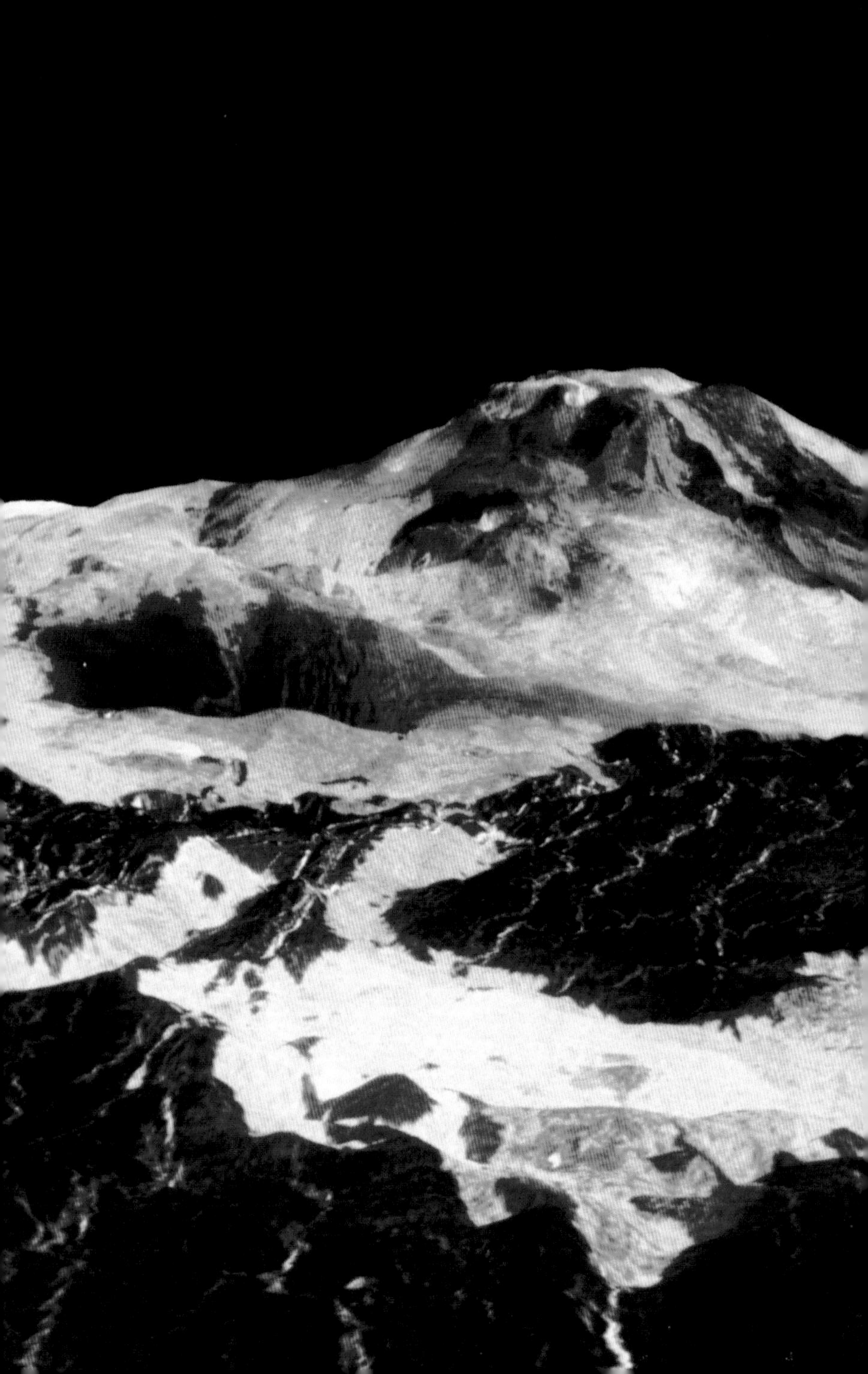

MAPPING THE SURFACE

In 1990 NASA's new spacecraft, Magellan, was able to reveal approximately 98% of Venus's surface. A total of 167 volcanoes with diameters greater than 100km were discovered, and many thousands of smaller volcanoes were observed. Venus was also shown to have a much smoother exterior than the Earth, as much of its surface was created by lava flows. The Venusian surface is thought to comprise mainly relatively young, basaltic rock, perhaps between 200 million and 800 million years old, indicating recent volcanic activity. On Earth, the surface rocks are all of different ages, while on Venus, the whole planet seems relatively young, and this has led scientists to believe that the volcanoes were not created by plate tectonics, as on Earth, but rather as a result of a cataclysmic, planet-wide eruption.

LEFT: Maat Mons, Venus. Computer-generated false-color 3D view of a Venusian landscape, looking towards Maat Mons. This 5km high volcano appears in the center of the horizon. Radar-bright lava flows dominate the surrounding scene. In the foreground (towards bottom right) is an impact crater.

TIMELINE

1824 - The German astronomer Johann Franz Encke calculated the Earth to Sun distance using data collected during the 1769 transit of Venus. His figure of 95,000,000 miles was used for several decades.

1932 - Carbon dioxide was discovered in the atmosphere of Venus by the astronomers Dunham and Adams.

1962 - US astronomer Carl Sagan became the first to calculate the effect of the Venusian atmosphere on the surface temperature of Venus.

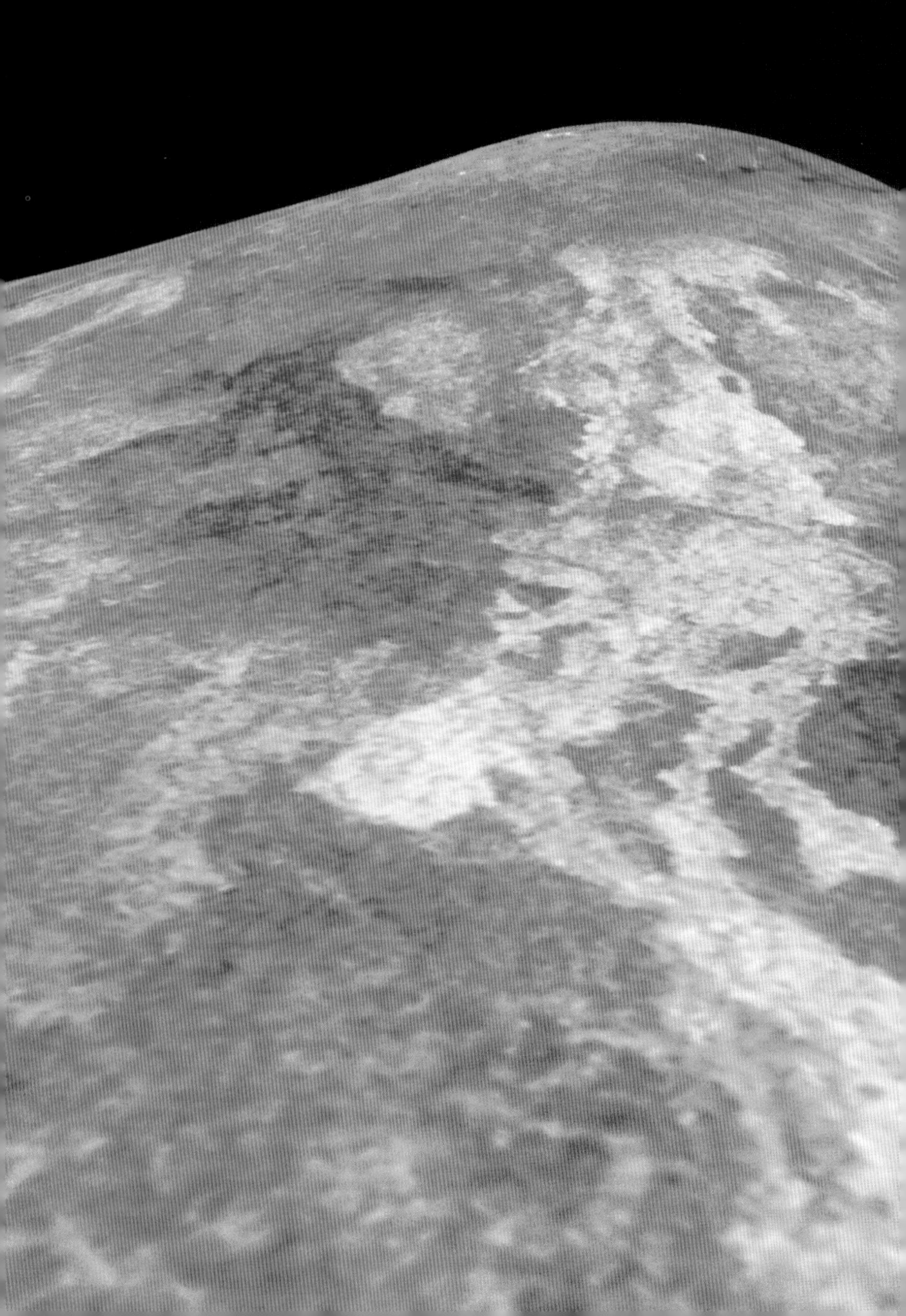

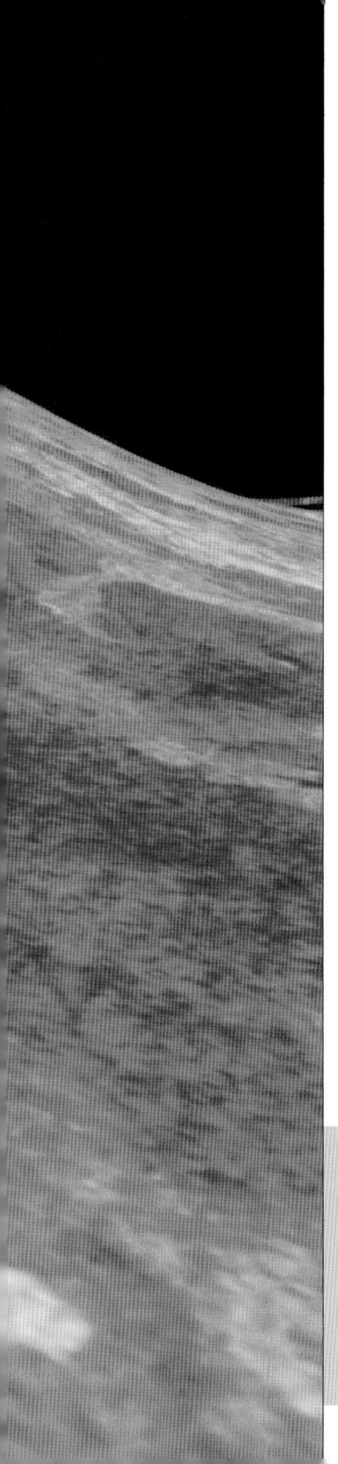

FUTURE MISSIONS

The European Space Agency launched its Venus Express at the end of 2005, for an arrival in early 2006. Once it reaches its destination, the Venus Express will measure surface temperatures from orbit, as well as try to give greater clarity to our understanding of the Venusian atmosphere and clouds. Japan also has plans to launch an orbiter, Planet-C, in 2007, to search for volcanic activity and lightning, as well as to continue analysis of the Venusian atmosphere. In the mid-1960s the gravity-assisted technique was developed. This enabled a spacecraft to gain a transfer of energy from a planet if it approached close enough, using the planet as a slingshot to catapult the craft to its destination in a much shorter time, with the use of less fuel. This gravitational slingshot idea has meant that Venus has been the indirect destination of many crafts, acting as a vital slingshot to propel them to the depths of the solar system. In a bid to make the probes less costly and more time-effective, most are given mini-missions to assess the planet that is catapulting them. Venus, as Earth's nearest neighbor, and the obvious contender to play slingshot, is a great beneficiary of this technique. It is not just missions to the superior planets which employ Venus as an accelerator, NASA's MESSENGER probe will require two Venusian flybys in 2006 and 2007 in order to reach Mercury, as will Europe's planned mission to Mercury, BepiColombo.

LEFT: Computer-generated topographical view of Sif Mons, a volcanic feature on the surface of Venus. Sif Mons is 2km high and 300km in diameter, located within the 2300km by 2000km rise in the western Eistla Regio region. A series of bright lava flows is seen spreading 120km from the peak.

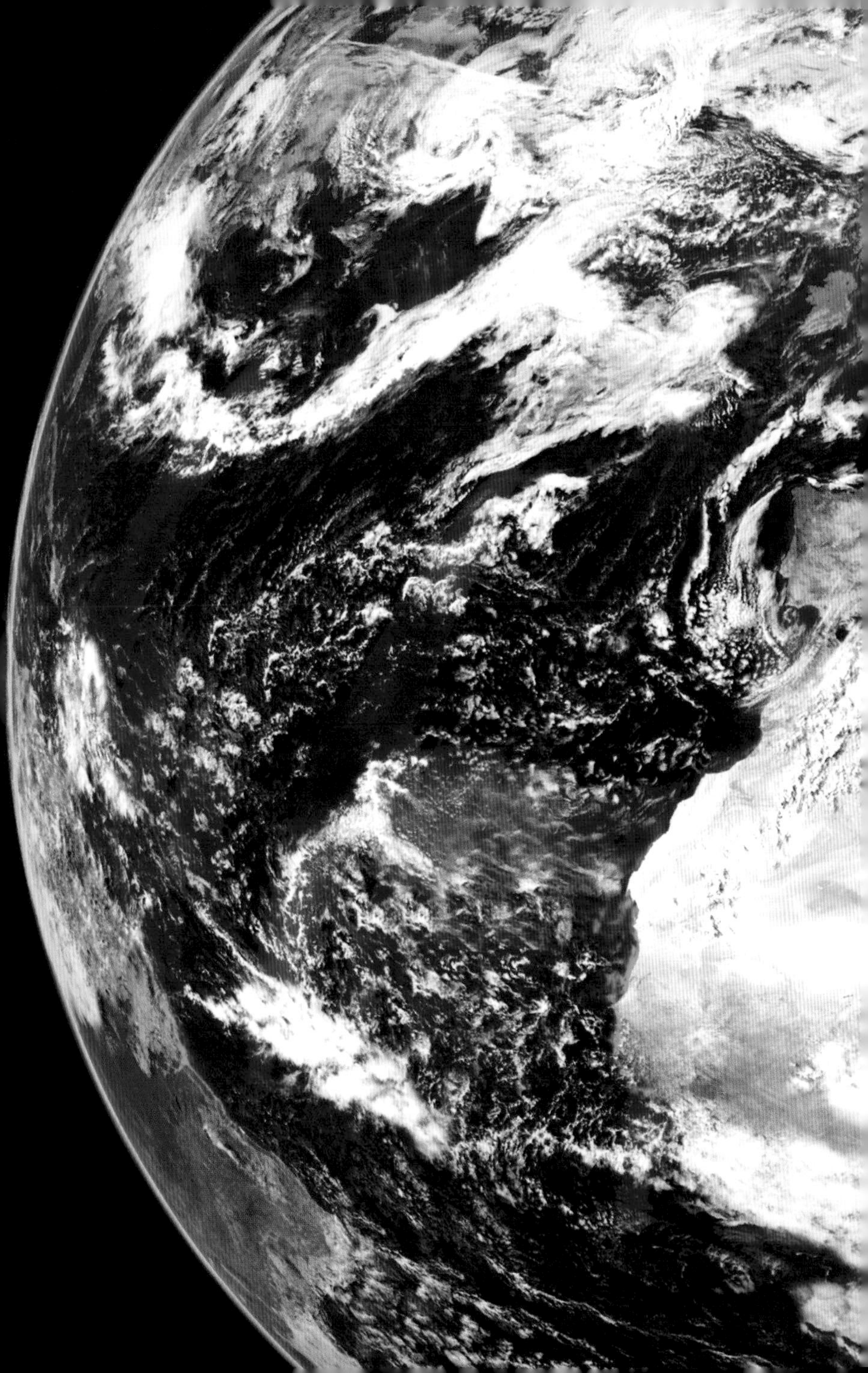

EARTH

THIRD ROCK FROM THE SUN

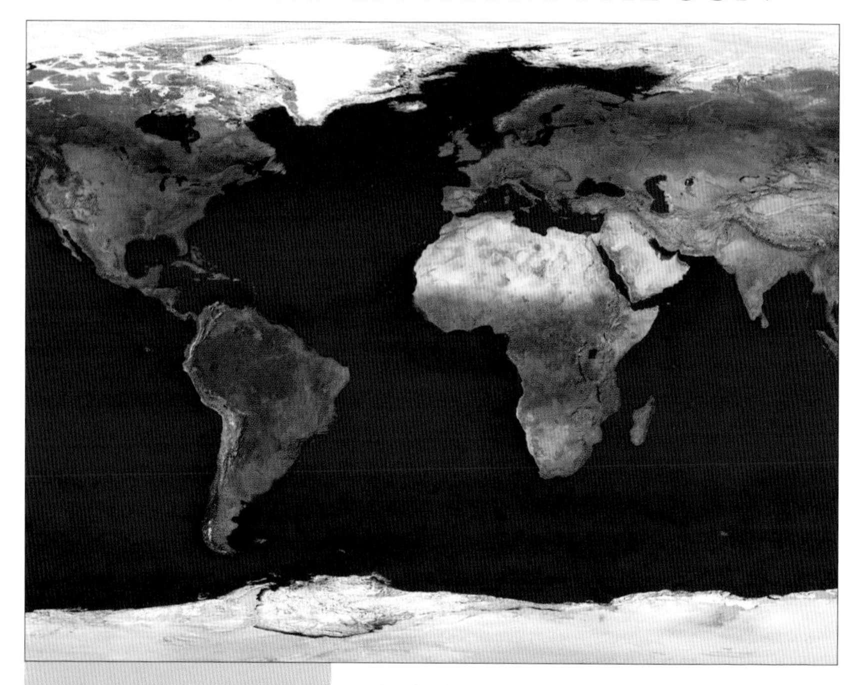

ABOVE: Whole Earth map based on satellite data. The land is dominated by vegetation or arid land, such as the deserts of North Africa. The polar regions are covered by snow and ice on sea and land.

PHYSICAL DATA

DISTANCE FROM SUN (AVERAGE): = 149,597,000 km

ORBITAL PERIOD: = 365.25days

ROTATIONAL PERIOD: = 23 hours 56 minutes

SPEED ORBITING THE SUN: = 30km/sec

The third "rock" from the sun is called Earth; it is also sometimes referred to as Terra, after the Roman goddess of the Earth, in keeping with a solar system named after characters from classical mythology. The Earth is unique in the solar system as the only planet with a nitrogen-oxygen atmosphere with liquid water oceans that cover more than 70% of the surface, and it is the only planet known to harbor life. The Earth remains geologically active; the surface is under constant, gradual change. It has a relatively large satellite, the moon, which is more than one quarter the size of the Earth. As a consequence of this size ratio the two objects are often called a double planet.

THE SEASONS

ABOVE: Crescent Earth. View of the Earth taken from the Apollo 4 spacecraft, in orbit 17,960km above the surface. The large amount of cloud cover makes it difficult to identify any surface features in the sunlit side.

The Earth takes 365.26 days to orbit the sun at an average distance of 150 million kilometers. It takes the planet 23.93 hours to revolve once on its axis. Earth's axis is not perpendicular to the planet's orbital plane, but is tilted 23.5 degrees. This tilt means that a particular line of latitude will receive a different quantity of solar heat as the Earth proceeds on its orbit. The seasons proceed from spring to summer to the fall to winter but climatic changes are more noticeable in the temperate and polar regions of the Earth. While most places experience longer hours of sunshine during summer and more hours of darkness in winter, the closeness of the polar regions to the poles means this seasonal change is exaggerated.

PHYSICAL DATA

TEMPERATURE RANGE: -90 to 60 degrees C

ATMOSPHERIC PRESSURE: = 1.03kg/cm^3

SURFACE GRAVITY: = 9.8m/s^2

DIAMETER: =12,756km

CIRCUMFERENCE: = 40,075km

SURFACE AREA: = 510,072km^2

TILT OF AXIS: = 23.5 degrees

MASS: = 5.97×10^{24}kg

VOLUME: = 1.1×10^{12}km^3

DENSITY: = 5.5g/cm^3 (the densest planet in our solar system)

BIOSPHERE

The Earth is the only planet we know to support life. Life appears to be resilient; every part of the planet plays host to some form. Life has been discovered at depths of 10,000 meters below sea-level and at heights in excess of 20,000 meters above sea-level. Life can be found in the ocean, in the skies, in the desert and in the polar regions but it is found in greatest abundance in the tropical regions. The fact that life proves so resilient on Earth indicates that it might be able to exist, in some form, elsewhere in the solar system, which is why many scientists have not abandoned hope of finding life there.

LEFT: The sun setting behind the limb of the Earth. Photographed by the Apollo 12 mission of 1969, returning from the second manned landing on the moon.

TIMELINE

1957 - Sputnik 1(USSR) became the first artificial satellite of the Earth.

1959 - Luna 1 (USSR) was the first successful mission to the moon and the first spacecraft to leave Earth's gravity.

1960 - NASA launched TIROS, the first weather satellite.

1961 - Vostok 1 (USSR) carried the first human, Yuri Gagarin into space. Alan Shepard became the first US astronaut in space.

1962 - John Glenn was the first American to orbit the Earth.

1964 - Nimbus 1 began a series of missions to study Earth's surfaces.

1968 - The first humans orbited the moon.

1969 - Apollo 11 became the first manned lunar landing.

1973 - Skylab, the first space station was launched.

1978 - The TOMS instrument launched on Nimbus VII recorded continuous data on Earth's ozone layer.

2000 - NASA launches first satellite of Earth Observing System (EOS) series, Terra.

ATMOSPHERE

ABOVE: Satellite image of the Earth, centered on the Arabian Peninsula. The Arabian Peninsula is separated from Africa by the Red Sea. It is also surrounded by Europe, the rest of Asia and the Indian Ocean

The Earth's atmosphere comprises approximately 78% nitrogen, 21% oxygen, and 1% argon, with trace amounts of other noble gases, hydrogen and carbon dioxide. The current composition is thought to be the third arrangement of the atmosphere since the creation of the Earth. The first would have consisted primarily of hydrogen gas together with some helium and would have evaporated into space relatively early in Earth's existence, when the planet was still molten. When the surface cooled sufficiently to crust over, volcanoes formed and Earth's gravity would have held the volcanic gases close by, forming a second atmosphere composed primarily of carbon dioxide and water vapor. The water vapor in the atmosphere would have rained down to form oceans, in which carbon compounds would provide the breeding ground for bacterial life and later plants, which would give the atmosphere its oxygen. The situation then became ripe for the nitrogen cycle to begin eventually resulting in atmospheric nitrogen.

ATMOSPHERIC LAYERS

The Earth's atmosphere is divided into a number of layers. Ascending from the surface to a height of 8km at the poles and 18km at the equator is the troposphere, named after the Greek word for mixing, because in this layer the atmosphere is circulated by winds. Conditions in the troposphere vary from region to region and from time to time; the state of the atmosphere in any given place at any given time is called weather. Weather is caused by differences in the amount of energy an area receives from the sun, which leads to variations in atmospheric pressure, temperature, humidity, wind and cloud cover. Above the troposphere is the stratosphere. It is heated by ultraviolet radiation, which is absorbed by ozone particles. High up in the stratosphere the ozone becomes more abundant and forms the ozone layer. In the outermost layer, the exosphere, atmospheric gases can escape into space. It extends far into space and many satellites orbit within it.

ABOVE: Sun setting behind the Earth as seen from the Space Shuttle Columbia. The tops of thunderclouds are seen towering high into the atmosphere. The glow of the sun is also seen reflecting off the upper layers of the atmosphere, providing a sharp dividing line between the Earth and space.

LAYERS OF THE ATMOSPHERE

TROPOSPHERE: up to 8 -18km above Earth's surface

STRATOSPHERE: up to 50km above Earth's surface

MESOSPHERE: from 50 - 85km above Earth's surface

THERMOSPHERE: from 85 - 450 km above Earth's surface

EXOSPHERE: from 450 - 10,000km above Earth's surface

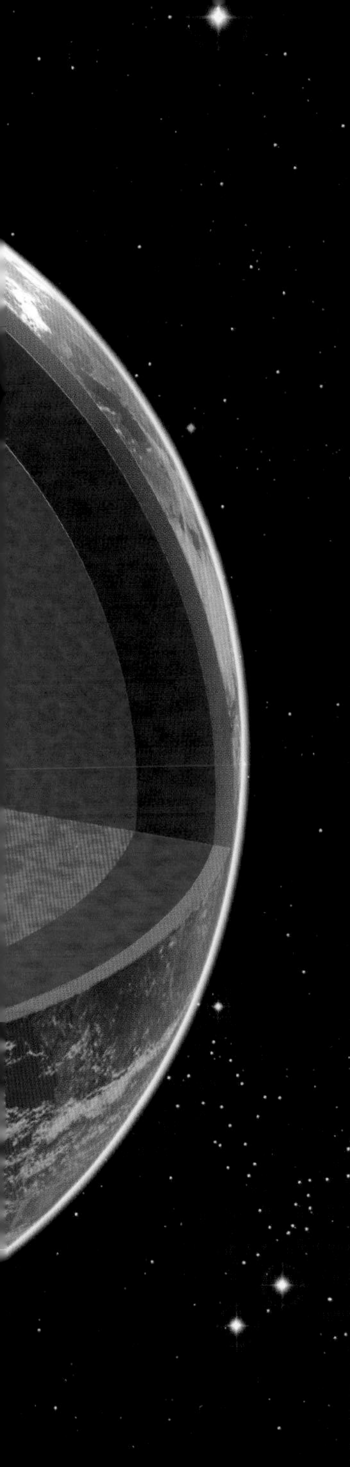

INTERNAL STRUCTURE

In its youth, the Earth was a ball of molten rock. As the planet cooled over time, the surface layer crusted over leaving molten rock beneath. The process of cooling has left the Earth in the state it is today. A solid iron inner core, more than 5000km below the surface, is enveloped by a molten iron-nickel outer core. Surrounding the outer core is the mantle, a layer of semi-molten rock of a viscous consistency ranging from 30km to almost 3000 kilometers in depth. Above the mantle is the Earth's crust, which averages 30km in thickness. The top of the crust is the surface layer of the land but much of it is sea-bed and therefore cannot be seen. However, the parts of the surface above sea-level carve out the familiar shapes of continents we recognize today. Although much of our information about the internal structure of the Earth comes from analysis of the behavior of shockwaves from earthquakes as they pass through the Earth, our understanding will be best clarified by going there. The deepest drills have barely scratched the surface of the crust but that will all change if a new Japanese mission to the upper mantle is successful in 2007.

LEFT: Cutaway computer artwork showing the internal structure of the Earth. From the center outwards, the five layers shown are: inner and outer core, inner and outer mantle, and crust.

PHYSICAL DATA

CORE: = 15% of volume of Earth; heavy iron and nickel. Inner core is solid; liquid outer core generates Earth's magnetic field

MANTLE: = 80% of volume of Earth; molten rock

INNER MANTLE: Temperatures exceed 6000 degrees C in the core

CRUST: = 5 - 70km layer of rock

CHANGING SURFACES

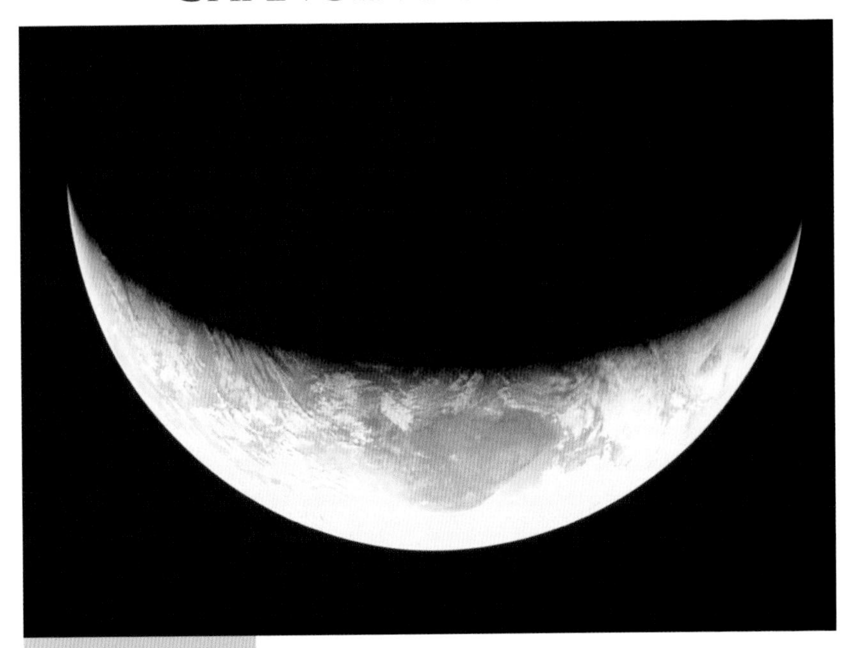

ABOVE: Crescent Earth, photographed in July 1969 from the Apollo 11 spacecraft, during its historic flight to the moon for the first manned lunar landing.

The surface of the Earth undergoes constant change because the crust is fractured into a number of plates. All the plates glide slowly over the upper mantle but not all move in the same direction. This means that plates are bound to collide, diverge or rub against one another as they move about and give rise to volcanoes, earthquakes or often both. If two plates are moving away from one another, chasms open up in the mantle below allowing hot magma to well up to the surface and form new land. This process usually occurs in the depths of the ocean and so is often termed "sea-floor spreading". This means that Europe and the Americas are gradually moving apart and is a reason for the changing face of the Earth over the years. There is only a finite amount of room on the surface of the Earth, meaning that when land is constructed at some plate margins, it must either be destroyed or elevated at another. If two continents collide, the result is that the land is pushed upward to form great mountain ranges. This is how the Himalayas were formed, when India crashed into the south of Asia.

PLATE TECTONICS

However, when a continental plate collides with an ocean-based plate, the result is destructive. Oceanic plates comprise basaltic rock, which is heavier than the mostly granite continental plates and so when they collide, the heavier oceanic plates are forced beneath the lighter continental plates into the mantle below, a process called subduction which results in the destruction of part of the crust. The process is slow; there may be no activity for many years while the plate builds up pressure. Eventually an almighty surge occurs as the oceanic plate charges into the mantle causing earthquakes and tidal waves. A tsunami in December 2004 killed an estimated quarter of a million people when the oceanic Indo-Australian plate was forced beneath the Burmese plate causing the displacement of water in the Indian Ocean. Water and other volatile materials, which accompany the plate into the mantle, cause a great deal of pressure, eventually resulting in a great explosion back through the crust in the form of a volcano. The Andes and the Rocky Mountain ranges were formed in such a fashion.

ABOVE: Popocatepetl. This volcano (5452 meters elevation) is the second-highest peak in Mexico. It had been dormant for around 60 years when it became active again in 1994. Moderate eruptions and earthquakes have occurred every few months, as of early 2001. As here, these can be accompanied by a column of ash and gas rising for several kilometers into the sky, posing a hazard to aviation. The largest eruption for 1000 years occurred in December 2000. Monitoring of the volcano and evacuation alerts help to protect the surrounding population. Photographed in December 1998.

MOON

OUR NATURAL SATELLITE

ABOVE: Crater Tsiolkovsky on the far side of the moon. The moon's orbit of Earth is locked so that only one hemisphere faces Earth at all times. The far side was seen for the first time when Russian probes photographed it in the late 1950s.

PHYSICAL DATA

AVERAGE DISTANCE FROM THE EARTH: = 384,000 km (Because of the gravitational pull of the Sun, the extreme ranges of the Moon from the Earth are from 356,400 km to 406,700 km)

ORBITAL PERIOD: = 27 days 7 hours 43 mins

ROTATIONAL PERIOD: = 27 days 7 hours 43 mins

The moon is tidally locked into a synchronous orbit of the Earth – it takes the same amount of time to rotate on its axis as it takes to orbit its parent planet. Therefore, during its 27 day, 7 hour and 43 minute rotation and orbit, the moon only ever presents one face to the Earth. The far side of the moon remained a mystery until 7 October 1959 when the Soviet probe Luna 3 first photographed it. The moon and the Earth are only 384,000 kilometers apart, meaning both bodies exact a strong gravitational pull upon one another, causing both bodies to bulge towards one another. This bulge is scarcely noticeable; however, the gravitational attraction causing it results in the tides we see on Earth's oceans.

FORMATION

It is speculated that the moon might originally have come from the Earth, the resulting debris broken off when Earth was impacted by a massive object early in our planet's history. Such an impact might have been responsible for the tilt of Earth's axis. The Earth and the moon also seem to be moving apart at a rate of an inch each year; if this process has continued throughout history, the Earth and the moon would have been much closer together in their past. Whether these two bodies were once one or not, analysis of rock from the moon suggests that the moon and the earth are very similar in age, and that they might even have been formed from the same nebula at the same time.

ABOVE: Photograph of the south-east portion of the moon. The dark patches are lava-filled basins called maria.

PHYSICAL DATA

AGE: Scientists believe that the Moon was formed approximately 4.5 billion years ago (the age of the oldest collected lunar rocks)

SURFACE AREA: = 37,932,330 km^2 (Earth = 510,065,700 km^2)

THE LUNAR SURFACE

It is not difficult for an observer from Earth to distinguish darker and lighter patches on the moon. The darker patches are named maria, the Latin for seas, because for generations it was believed that the darker patches were likely to be oceans. These "seas" are smooth, lowland areas comprising basalt, which are the result of lava flows. Over three billion years ago, large meteor impacts would have exposed molten rock beneath the crust, which would have flowed out, forming the lava plains we see today. Perhaps the most famous of all the maria is the Sea of Tranquillity, famed for playing host to the first moon landing. Surrounding the maria are the highland areas which are far from smooth, rather they are pockmarked by craters, the scars of a long history of impacts with comets and meteors. The ray-shaped crater Tycho is particularly noticeable from Earth as, from its position in the south central section of the moon, it reflects light from the sun. Another prominent crater is Copernicus, and, although it is not as impressive as Tycho, it can be observed clearly with binoculars. The highland areas are covered in regolith, a fine rock and dust coating, which are the result of consecutive meteors breaking up upon impact and spreading their debris across the surface. The far side of the moon has far fewer maria than the near side, instead it appears pockmarked by years of meteor impacts, which did not cause bleeding of molten rock from the mantle onto the surface; this indicates that the crust is thicker on the far side of the moon than it is on the near side of the moon, which has considerably more maria.

LEFT: Earthrise. This photograph of a blue and white Earth rising over the horizon of the moon was taken from the Apollo 11 spacecraft in July 1969. The lunar terrain in the field of view is in the area of Smith's Sea.

MISSIONS TO THE MOON

ABOVE: Apollo 11 astronaut Edwin "Buzz" Aldrin is photographed on the surface of the Moon, next to the United States flag. Part of the lunar module is seen on the left of the photograph. Apollo 11 was the first manned lunar landing mission and was launched on 16 July 1969. It landed on the moon on 20 July 1969.

LUNAR MISSIONS FROM OTHER COUNTRIES

1990 - Hiten (Japan) was the first Japanese lunar mission; it made a total of 10 lunar orbits before the spacecraft was deliberately crashed into the Moon's surface

1997 - AsiaSat 3/HGS-1 (China); AsiaSat 3 was designed as a communication satellite, but after a rocket malfunction failed to put it into Earth's orbit, pioneering techniques were used to send the spacecraft to the Moon in order to stabilize its orbit round the earth. The satellite was re-named HGS 1

As mankind entered the age of space travel, the moon, as the largest and brightest body in the night sky, became the obvious target for both the United States and the Soviet Union. In 1959, the Soviet probe, Luna 2, crashed onto the surface of the moon and became the first man-made structure to land on the surface of another world. A month later, another Soviet probe, Luna 3, sent back the first-ever pictures of the far side of the moon. In 1962, President Kennedy announced that the United States would send a man to the moon by the end of the decade. America's lunar program did not get off to a successful start, its three Able space probes had failed to reach the moon in 1958. But by 1964, the United States crashed its first probe onto the moon. Ranger 7 took the first close-up pictures of the moon's surface as it crashed in July 1964 and consecutive American and Soviet missions to the moon deepened our understanding of Earth's satellite.

BACTERIAL CONTAMINATION

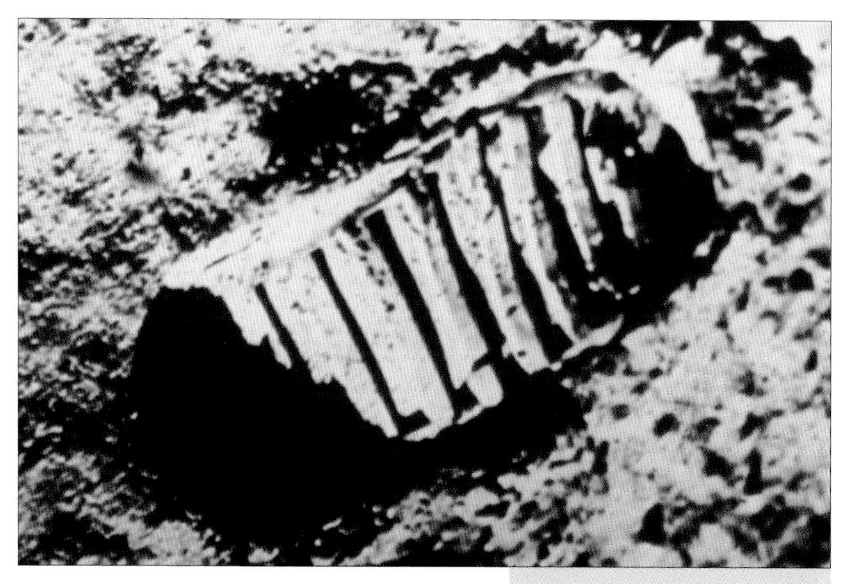

In 1966 the Soviet Probe, Luna 10, became the moon's first artificial satellite. Its greatest achievements were identifying and monitoring the Moon's weak magnetosphere as well as detailing lunar radiation. The same year, American probe Lunar Orbiter 1 took the first pictures of the Earth as seen from the moon and the following year, in 1967, successfully landed a probe, Surveyor 3, on the moon's surface; it was equipped with a mechanical scoop to analyze the lunar soil. When the Apollo 12 crew picked up the Surveyor 3 probe, three years later, it was discovered that bacteria, which had got into the probe before its launch, had been able to survive on the moon for three years. As a result subsequent probes are fully sterilized to avoid contaminating their destination.

ABOVE: The footprint of Neil A. Armstrong's first step on the moon on 20 July 1969.

EARLY MISSIONS

1959 - Luna 1 (USSR): Most of Russia's unmanned Moon missions were undertaken by the Luna program. 17 of the 45 Luna missions were successful. Luna 1 was the first spacecraft to flyby the Moon and the first to go into orbit around our Sun

1959 - The early Pioneer missions marked America's first efforts to reach the Moon. All eight Pioneer moon shots provided important information for future space programs but Pioneer 4 was the only one of the series to achieve its goal. It was the first successful mission to the Moon, the first US spacecraft to escape Earth's gravity and the first American spacecraft to achieve an orbit around our Sun

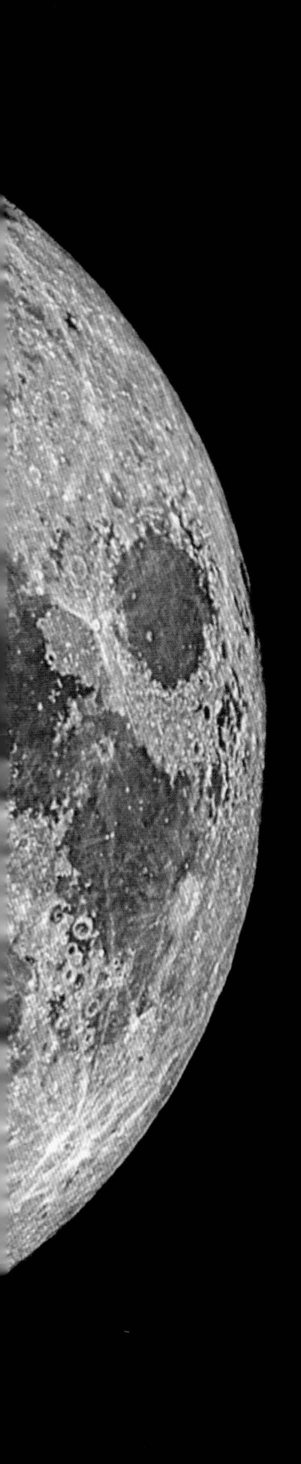

APOLLO 11

In December 1968, the United States successfully sent a manned mission, Apollo 8, into lunar orbit. This was closely followed by people across the world, and whetted appetites for more. On 16 July 1969, Apollo 11 blasted off from the Kennedy Space Center in Florida. Aboard, the three astronauts, Edwin "Buzz" Aldrin, Neil Armstrong and Michael Collins, accompanied the craft on one and a half orbits of the Earth and then its journey to the moon. On 20 July, the lunar module, Eagle, separated from the command module, Columbia. Michael Collins remained on Columbia, while Aldrin and Armstrong descended to the moon's surface in the Eagle module. As an estimated 500 million people around the world watched in amazement, the module landed successfully on the moon as Neil Armstrong uttered the first of his two iconic quotes of the day, "The Eagle has landed." Just before 9pm – the time in mission control, Houston, Texas – Neil Armstrong stepped out of the probe onto the powdery surface of the moon and uttered perhaps the most famous quote of the twentieth century, "That is one small step for man, one giant leap for mankind."

LEFT: Optical image of a full moon 14 days into its 28-day cycle. The moon is full when the Earth-facing side is fully lit by the sun. The Earth-facing side of the moon is locked in place by the gravitational influence of the Earth. This results in a lunar day that equals the length of the lunar Earth orbit.

THE APOLLO PROGRAM

ABOVE: Apollo 11 Lunar Module returning from the surface of the moon to dock with the Command Service Module, while in the background the Earth rises above the horizon of the moon.

PHYSICAL DATA

MASS: = 7.35×10^{22} kg (Earth = 5.97×10^{24} kg) The ratio of the mass of the Moon to Earth is far larger than the similar ratios of other natural satellites to the planets they orbit, with the exception of Charon and Pluto

VOLUME: = 2.2×10^{7} km^3 (Earth = 1.1×10^{12} km^3) If the moon were seen next to the Earth, it would look like a tennis ball next to a football

DENSITY: = 3.3 g/cm^3 (Earth = 5.5 g/cm^3)

Before the end of the 1960s America returned to the moon; Apollo 12's lunar module landed on the moon on 18 November 1969. Apollo 12 landed in a different sea to Apollo 11, the Ocean of Storms, while the first moon landing had been in the Sea of Tranquillity. Rocks returned to Earth by both missions revealed that the maria had formed at different periods. This gave scientists the first indications that the maria had been formed by lava bleeding onto the surface after severe meteor impacts. Over the next three years NASA sent a further five Apollo missions to the moon, the most memorable being the mission of Apollo 13. In April 1970, an oxygen tank exploded while Apollo 13 was making its way towards the moon. The moon landing was, of course, cancelled and remarkably all three astronauts returned to Earth safely.

MOON BUGGIES

Apollo 14 in January 1971 resumed the successful moon landing program and one of the crew, Alan Shepard, famously practiced his golf swing on the moon. Apollo 14 was followed by three further successful missions, Apollos 15, 16 and 17. These last three missions all included a lunar rover, also known as a "moon buggy." This vastly increased the area the astronauts could cover; astronauts from the Apollo 17 mission covered over 30km of ground in their lunar rover and astronauts from Apollo 16 hold the record for the fastest speed for a wheeled vehicle on the moon: 11 miles per hour. Three further Apollo missions were planned but Apollos 18, 19 and 20 never got off the ground, and Apollo 17 became the last manned mission to the moon to date.

ABOVE: Lunar landing module. The lander Challenger in the Taurus-Littrow valley on the surface of the moon during the Apollo 17 mission of 1972.

PHYSICAL DATA

SURFACE GRAVITY: = 1.6 m/s^2 (Earth = 9.8 m/s^2) The gravitational forces between the Earth and the Moon affect the level of the ocean tides, causing the Earth to have two high tides per day. If you weigh 100 kg on Earth you would weigh16 kg on the Moon

TEMPERATURE RANGE: -233° to 123° C (Earth = -90° to 60° C)

LUNAR PHASES

ABOVE: Image of a waxing crescent moon 5 days into its 28-day cycle. The lunar phases arise as the moon's orbit of the Earth shows the Earth-facing side moving into and out of the light of the sun.

The moon undergoes phases, changing from a full moon to a crescent, half or gibbous moon, or even no moon at all – a new moon. These phases are caused by the changing angle at which the illuminated face of the moon can be observed from Earth. Such changing angles occur because the moon is moving around the Earth, but only presenting one face to the Earth. Therefore, whenever a full moon cannot be seen, it means that a portion of the far side of the moon is illuminated, but we cannot see the far side from Earth. After 27.3 days the moon returns to its original position and the cycle begins again.

WAXING AND WANING

The cycle begins with a new moon. Between the new moon and the full moon, the moon is said to be waxing because the portion of the near side of the moon being lit up by the sun is growing. Halfway between the full moon and the new moon, when the moon intercepts the orbital path of the Earth, we see a half moon, because half of the near side is illuminated. Between the new moon and the half moon we see a crescent moon, as only a sliver of the moon is revealed to us. Between a half moon and full moon, we see a gibbous moon. Halfway through the moon's orbit, we see a full moon, as the moon moves behind the Earth, so that the whole of the near side of the moon is illuminated, but we cannot see the far side from Earth.

ABOVE: Composite image showing the moon at each stage of its 28-day cycle. Along the top row, the moon is a waxing crescent, reaching a half moon after 7 days. For the next seven days it is a waxing gibbous, reaching a full moon after 14 days. The moon is then a waning gibbous,reaching another half moon 21 days into the cycle. On the bottom row it is a waning crescent, reaching the new moon stage at bottom right.

ECLIPSES

When there is a new moon, the moon lies between the sun and the Earth, but cannot be seen because only the far side, unseen from Earth, is lit up. New moons occur once every 27.3 days, and usually the moon passes above or below the ecliptic plane, the plane upon which the Earth and the sun lie. However, sometimes the new moon enters the ecliptic plane and obscures the sun; this is called a solar eclipse. When it is closest to the Earth, the moon appears larger in the sky, and therefore blocks out the entire disc of the sun, causing a total solar eclipse. When the moon is farthest away from the Earth it is not large enough to block the disc of the sun entirely; here we get an annular eclipse, and a ring of the sun can be seen encircling the dark disc of the moon. Total solar eclipses are not actually that rare, they occur somewhere on Earth every one to two years, but it is rare for a recurrence in the same spot on the Earth's surface for several decades.

If the moon enters the ecliptic plane on the far side of the Earth, the Earth casts a shadow onto the moon; this is a lunar eclipse. When the moon enters the umbra of the Earth's shadow, it does not usually disappear from sight, because sunlight is refracted by Earth's atmosphere onto the moon, and the moon shines with a reddish hue. Lunar eclipses are not uncommon, at least two occur every year; however, they are not always so spectacularly colored; sometimes the refraction of sunlight might be obstructed by cloud or dust and the moon will appear much darker.

LEFT: Eclipsed lunar disc. Image of a partial phase of a lunar eclipse at totality. Lunar eclipses are caused by the entry of the moon into the cone of the shadow cast by the Earth.

FOLLOWING PAGES: Lunar landscape. Basalt rocks littering the south rim of the Camelot crater, in the valley of Taurus-Littrow, on the surface of the moon.

FROM THE PAST

ABOVE: Crescent moon with Earthshine. The moon does not produce its own light, but can be seen because it reflects light. The bright crescent is lit by the sun whilst the darker part is lit by "Earthshine" or sunlight reflected from the Earth.

OPPOSITE ABOVE: Waning gibbous moon 19 days into its 28-day cycle. The lunar phases arise as the moon's orbit of the Earth shows the Earth-facing side moving into and out of the light of the sun.

TIMELINE

1610 Galileo made the first telescopic observation of the Moon.

1835 - The New York Sun wrote a hoax report that William Herschel had discovered life on the moon. It caused widespread excitement as many believed the report, demonstrating that popular belief in lunar life had not been widely extinguished with the development of the telescope.

The moon, as the largest object visible in the night sky, has understandably captivated spectators on Earth since the dawn of humanity. In many cultures the moon had important religious connotations and therefore scientific understanding of the moon made slow progress over the centuries. In ancient Greece, Anaxagoras was imprisoned for suggesting that the moon, as well as other celestial bodies, was spherical. Since ancient times it had been thought that the dark patches on the moon might be large seas of water and that the moon harbored life. However, with the arrival of the telescope in the seventeenth century, the moon was shown to be a dry, cratered world, seemingly incapable of supporting life, or holding liquid water.

TO THE FUTURE

Apollo 17 became the last manned lunar mission; the United States government did not believe they could justify further funding when so much data had been gathered already and the country was fighting a costly war in Vietnam. Moreover, the Apollo program had not resulted in any fatalities, even Apollo 13 had got back to the Earth safely. By quitting while it was ahead, NASA ensured it maintained a clean sheet in terms of returning its astronauts to Earth. In 2004, US President, George W. Bush, announced that America would return to the moon by 2020. The last time a President had set a date on a manned lunar mission, the USA met the challenge and successfully landed the first man on the moon before the time expired. Even if President Bush's timeframe is not met, it is a reliable bet that humans will once again walk upon the surface of the moon.

PROPOSED FUTURE MISSIONS

Japan's LUNAR-A has been designed to penetrate the lunar surface to study the Moon's interior with seismometers and heat-flow probes

SELENE (SELenological and ENgineering Explorer) is a lunar exploration mission to be conducted by the Institute of Space and Astronautical Science (ISAS) and the National Space Development Agency (NASDA) – Japan's two major space agencies. The primary objectives of the mission are to obtain data on the Moon's origins and evolution

Lunar Reconaissance Orbiter. The proposed Lunar Reconaissance Orbiter is a Moon orbiting mission scheduled for launch in 2008. A primary goal of the mission is to find and identify landing sites for future robotic and human explorers

MARS

THE RED PLANET

ABOVE: Artwork of Mars as it is today: a dry, barren planet. It is thought that liquid water existed on Mars early in its history, but was lost to space over time. This may have been due to Mars's weak gravity, thin atmosphere and weakening magnetic field. If liquid water did exist on Mars, there is a chance that life may have begun there.

PHYSICAL DATA

MASS: = 6.42×10^{23} kg (Mars is about 10% as massive as the Earth)

VOLUME: = 1.6×10^{11} km^3 (about 15% of Earth's volume)

TEMPERATURE RANGE: = -143° to 20°C (Earth temperature range = -90° to 60°C)

Mars used to be referred to as the "dead planet" following Mariner 9's 1971 flying visit, from which the first pictures of the Martian surface revealed a dusty, moon-like, cratered world, devoid of other discernible features. This changed in the later 1970s, when the Viking missions reached Mars and revealed super-volcanoes, gigantic valleys and areas that appeared to be ancient, giant flood plains. The question of whether Mars is dead does however remain pertinent. Scientists remain unclear about the possibilities of life, water and volcanic activity on Mars. Owing to its small size and mass, the Martian atmosphere is exceptionally thin, approximately one hundredth the density of Earth's atmosphere. The chief component of the Martian air is carbon dioxide, which comprises around 95% of the atmosphere. The remaining 5% mainly consists of argon and nitrogen, with traces of oxygen.

WEATHER SYSTEMS

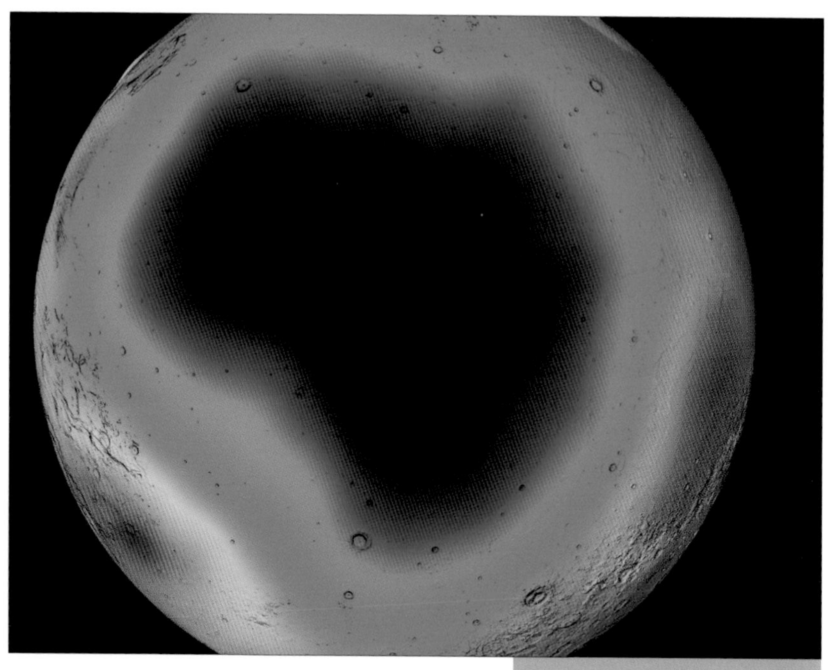

Mars is a cold planet. The average surface temperature is minus 60 degrees Celsius. In the summer months, the equator can reach temperatures of around 20 degrees Celsius. Such temperatures cause low pressure, and cold air rushes in from the poles to fill the gap left behind as this relatively hot air rises. This creates winds on Mars which can reach speeds of several hundred miles per hour. These fast Martian winds pick up dust particles and engulf Mars in massive, planet-wide dust-storms. When probes have taken color photographs of Mars, the sky appears a pinkish-tan color. This is the result of the iron-rich dust particles, which are whipped up from the surface by Martian winds, and are almost always present in the atmosphere. However, when the dust subsides the Martian sky changes to a dark blue.

ABOVE: Water ice on Mars. During the Martian summer, a large expanse of frozen water is seen on the surface of Mars. During winter, much of the ice is obscured by layers of carbon dioxide frost or snow. This image shows the northern hemisphere of Mars.

TIMELINE

1609 - Galileo first observed Mars.

1672 - Huygens observed a white spot at the south pole of Mars.

1704 - Italian astronomer Giacomo Filippo Maraldi (Cassini's nephew) observed white spots at the poles.

1927 - William Weber Coblentz and Carl Otto Lampland measured a wide temperature difference between the day and night side of Mars which suggested a very thin atmosphere.

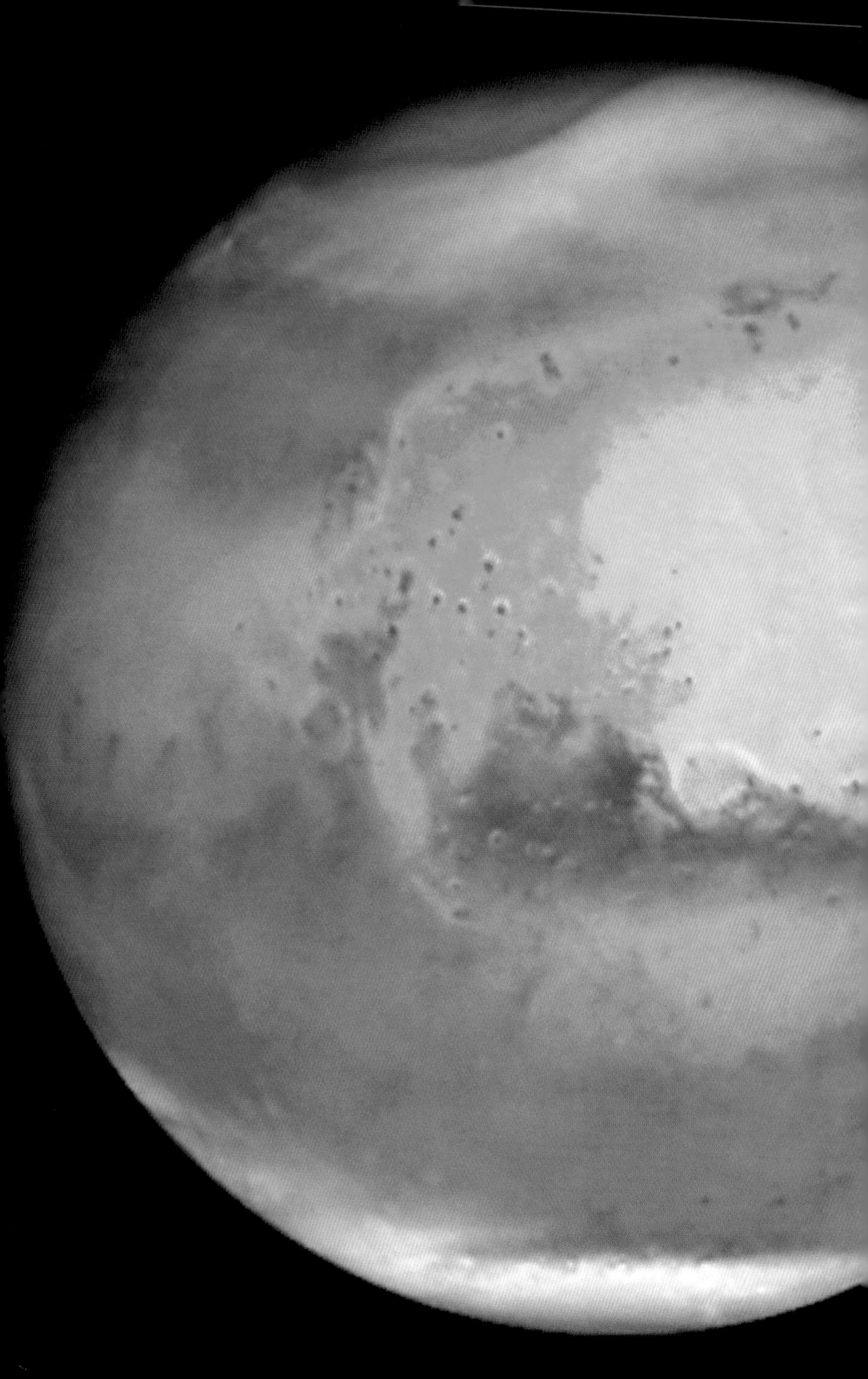

THE SEARCH FOR WATER

Water exists on Mars, in both frozen and vapor form, but it has not yet been found as liquid. This is because the surface of Mars is far too cold. Therefore, surface water is only to be found in frozen form and this is largely confined to the northern polar ice cap. Both poles have a residual cap that exists all year round, and both exhibit seasonal ice caps which can be found during the winter months.

The concept of liquid water on Mars is age-old, perhaps because the ice caps could be observed through telescopes. Telescopes had also picked up dark lines across the surface, which popularized the nineteenth-century idea that canals existed on the planet. When Mariner 4 reached Mars in 1964 it revealed a dead, cratered surface, putting paid to the idea of liquid water. But since Mariner 4 the case in favor of liquid water has been mounting; NASA's Viking missions, a decade later, detected the existence of fluvial valleys and channels in the northern hemisphere, and such evidence was bolstered by NASA's Pathfinder mission in 1997 which suggested that the layout of rocks in a region called Ares Vallis indicated they had been formed by a flood. NASA remains unwilling to unequivocally announce the existence of a liquid, watery past on Mars, but the evidence is becoming increasingly convincing.

LEFT: An optical image of Mars. The Martian summer in the northern hemisphere results in a large south polar ice cap. Two seasonal dust storms can be seen, one at top center and one over the Hellas impact basin at lower right.

TIMELINE

1719 - Maraldi suggested that the white spots on Mars could be ice caps.

1784 - Sir William Herschel observed the seasonal changes to the polar caps and raised the possibility that they might be composed of snow and ice.

1862 - British astronomer Sir Norman Lockyear agreed with earlier suggestions that there were oceanic areas on Mars.

LIQUID WATER

With favorable evidence of a liquid-water past, scientists are faced with two questions. Firstly: how did the water flow? The volcanoes on Mars could have created a greenhouse effect by emitting carbon dioxide. As a result, the atmosphere could have warmed, melting frozen water which would then have flowed in liquid form across the surface of the planet. Alternatively, if Mars has always been as cold as it is today, oceans of water could have flowed under a vast covering of ice, melted by friction as the ice moved across the surface of the planet.

The second question is: where did all the liquid water go? More recent missions to Mars have hoped to make sense of this conundrum. Convincing explanations include evaporation into space or that liquid water subsequently froze to form the ice caps we see today. A more recent theory is that a large amount of water might be locked in the sand dunes, cementing them together, explaining why they have been so durable over the years, and have resisted the fierce wind and dust. In 2002, NASA's Odyssey probe gave the most powerful indications yet of the presence of liquid water on Mars; it gathered evidence suggesting that vast amounts of water might be stored a few feet beneath the surface. If this proves accurate, it would not only re-ignite the possibility of manned settlements on Mars in the distant future, but it would also give scientists a clearer indication of where to search for fossilized evidence of Martian life forms.

LEFT: The dry, rocky surface of Mars, taken by the panoramic camera on the Mars Exploration Rover Spirit. NASA scientists chose this location because the 145-kilometer-wide crater may once have contained a lake.

PHYSICAL DATA

CIRCUMFERENCE: = 21,344 km (Earth = 40,075 km)

DIAMETER: = 6,796 km (about half the size of Earth)

ATMOSPHERIC PRESSURE: 0.007 kg/ cm^2 (This low pressure means a person would survive no more than a few seconds without a pressure suit)

THE SEARCH FOR LIFE

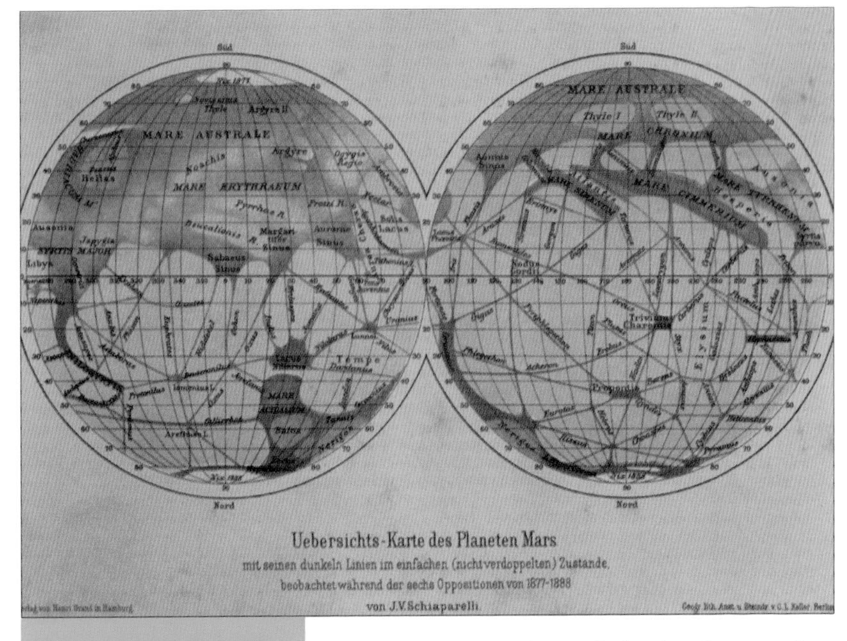

ABOVE: Historical map of the surface of Mars. This drawing was made by the Italian astronomer Giovanni Schiaparelli between 1877 and 1888. Schiaparelli called the straight surface features "caneli" and also noticed that the patterns on the surface changed with the Martian seasons.

PHYSICAL DATA

SURFACE GRAVITY: = 3.7 m/s (A 100kg object on Earth would weigh 38 kg on Mars)

DENSITY: = 3.9 g/cm^3 (70% as dense as Earth)

SURFACE AREA: = 144,100,100km^2 (Earth = 510,072,000km^2)

In 1877 Giovanni Schiaparelli developed a map of Mars, from observations through a telescope at his observatory near Milan. Schiaparelli's map displays lines criss-crossing Mars, which he termed "canali," the Italian for channels. However this term was often misinterpreted as "canals" which was taken to indicate the presence of water and also implied that someone had built these waterways.

The idea of intelligent life on Mars was not new. In the Middle Ages it was often taken as a given that all the planets were inhabited, even the sun. However, the possibility of "canali" focused attention on the idea of life on Mars. On many occasions, NASA has risen to the challenge of searching for life on Mars. Several of the Lander missions were dispatched with a mandate to see whether Mars was ever home to some form of life, and even to uncover whether microscopic life forms have endured in the Martian soil.

EVIDENCE OF LIFE

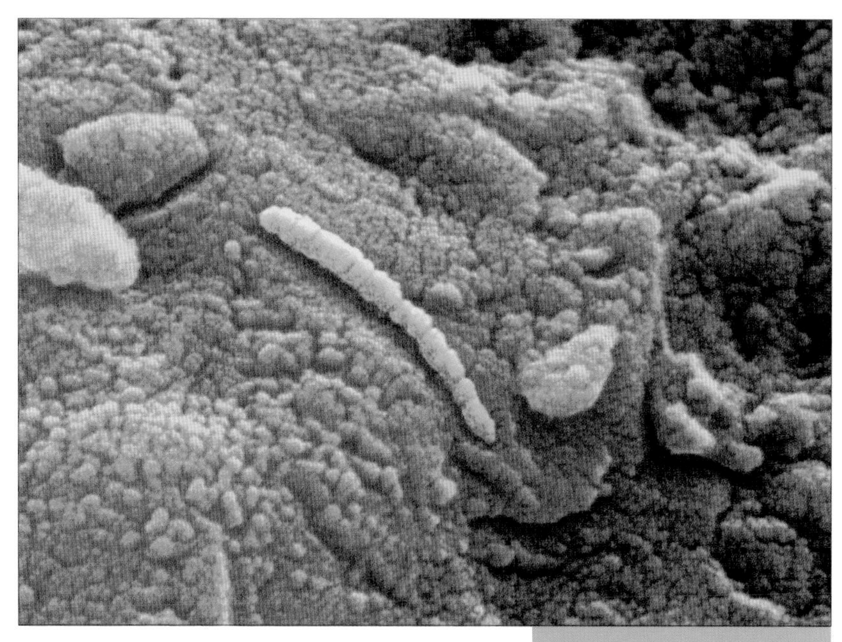

The debate as to whether there is life on Mars escalated in 1996 when NASA's analysis of a meteorite from the planet was released. The meteorite, ALH84001, was discovered in Antarctica in 1984, and it is estimated that the 4.5 billion-year-old rock fell to earth 13,000 years ago. The rock contained microscopic structures similar to bacteria on Earth. Such evidence points to the existence of life on Mars, but critics claim that the life forms are merely fossil microbes from Earth which infiltrated the rock after it landed. However, this is refuted by some scientists who point out that the life forms are only found at the center of the rock and not at its crust, as would be expected if they had permeated the rock after landing. NASA intends to move closer to a more conclusive answer as to whether there is life on Mars by sending Phoenix in 2007 followed by the Mars Science Laboratory in 2009.

ABOVE: Evidence for life on Mars. Colored scanning electron micrograph of a tube-like structure on a meteorite which originated from Mars. Structures such as this have been interpreted as possibly being microfossils of primitive, bacteria-like organisms which may have lived on Mars more than 3.6 billion years ago. The structures are less than 1/100th the diameter of a human hair.

TIMELINE

1698 - Huygens published Cosmotheoros and addressed the question of life on Mars.

1938 - When "War of the Worlds" (H.G. Wells) was broadcast it was estimated that over 1 million listeners thought the events were real!

THE MOONS OF MARS

ABOVE: A Mars Express image showing Phobos, the larger of the two Martian moons. Phobos is irregular in shape, measuring 19x21x27km. The large Stickney Crater (center left) is 10km across.

PHYSICAL DATA OF MOONS

PHOBOS

MEAN DISTANCE FROM MARS: = 9,377km

DIAMETER: = 21 km

MASS: = 1.06×10^{16}kg

MEAN DENSITY: = 1.90g/cm^3

DEIMOS

MEAN DISTANCE FROM MARS: = 23,459km

DIAMETER: = 12 km

MASS: = 2.38×10^{15}kg

MEAN DENSITY:= 2.2g/cm^3

Mars has two moons named Phobos, meaning fear, and Deimos, meaning dread, after characters in Homer's Iliad. Both moons were discovered by the American astronomer Asaph Hall in August 1877. They were undiscovered before because they are both so small. It is most likely that they strayed into Mars's gravitational pull from the asteroid belt.

Phobos is the largest of the two. It orbits Mars more closely than any other moon orbits its parent planet, meaning that Phobos creeps ever closer to the Roche Limit – the area within which a moon will break up due to gravitational tidal forces exerted upon it by the planet. When this happens, several million years in the future, the debris from Phobos will turn into a planetary ring around Mars. Meanwhile, Phobos orbits Mars faster than it rotates on its own axis, so it rises and sets twice during one Martian day, while it takes Deimos just over two Martian days. It becomes apparent how quickly both moons orbit Mars when they are contrasted with our moon's 28-day revolution period.

THE SURFACE

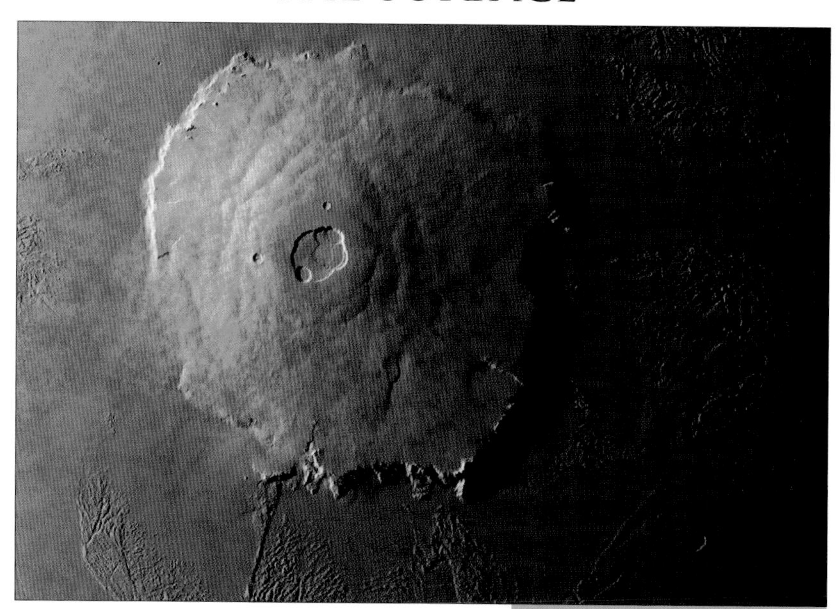

When Mariner 9 reached the planet in 1971, Mars was in the midst of a dust storm. As the dust subsided, the first image seen by scientists showed the tops of four massive volcanoes: Olympus Mons, Arsia Mons, Pavonis Mons and Ascraeus Mons. It was quickly realized that Olympus Mons was the biggest volcano in the solar system, towering over Earth's largest, Mauna Loa. It is seventeen miles high and 350 miles in diameter, eleven miles higher and 275 miles wider than Mauna Loa.

Scientists monitoring Olympus Mons have suggested that it last erupted between 20 and 200 million years ago. However, the flows seem to have been very small, implying that volcanic activity on Mars might be running out of steam as the mantle cools. If this is the case, it would be a shame that the planet has begun to burn out, just as humanity begins traveling there.

ABOVE: Artwork of Olympus Mons, the largest known volcano in the solar system. Its summit rises 27km above the surrounding plains (Mount Everest is 8.8km high), while the base measures 600km across, resulting in a gentle surface slope. The rough, crinkly patches around the volcano form the Olympus Mons Aureole.

MARS FACT

In 1840 Wilhelm Beer and Johann Madler created the first global maps of Mars.

MISSIONS TO MARS

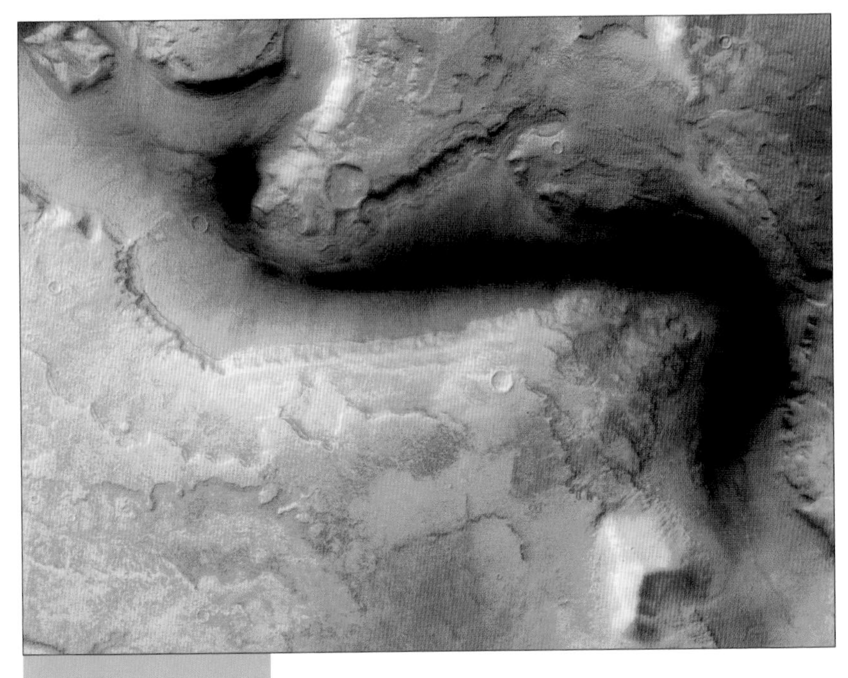

ABOVE: The surface of Mars, showing the Reull Vallis channel, which was created at a time in the the past when water flowed on the planet. The area is to the east of the Hellas Basin and measures 100km across.

TIMELINE

1659 - Christiaan Huygens made the first sketch of Mars and arrived at an approximate 24-hour rotational period.

1840 - Wilhelm Beer and Johann Madler created the first global maps of Mars.

Mariner 4, the first successful mission to Mars was launched in November 1964 and reached the planet in mid-July 1965. The probe traversed Mars at great speed, managing to snap photographs of the Martian surface. The pictures Mariner 4 returned finally ruled against the existence of canals on Mars. Craters, and not canals, covered the Martian surface. As a result, Mars became known as the dead planet; it bore more similarity to the Moon than to the Earth.

Soon after the moon landing, in summer 1969, the American probes Mariners 6 and 7 reached the Martian equatorial region and the southern polar region, respectively; their photographs still showed Mars as a dead planet riddled with craters. Finally, when Mariner 9 reached Mars in 1971, the speculation that Mars was a dead planet proved unfounded; the planet had supervolcanoes and giant valleys, dwarfing anything here on Earth. The probes before it had been unfortunate in looking at the regions with a disproportionately large number of craters.

MISSIONS TO MARS

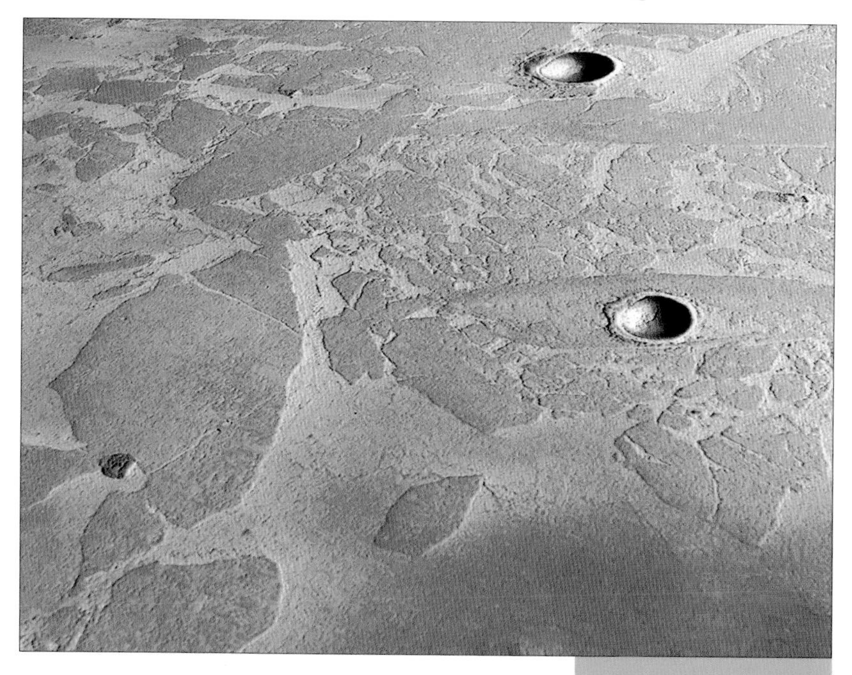

The USSR struggled to catch up; its first probe, Mars 1 had been lost in 1962 and in a bid to outdo its rival, dispatched two probes, Mars 2 and Mars 3, designed to land on the planet, in 1971. But Mars 2 did not relay any data from the surface and Mars 3 only sent back blurred images for just a few minutes before shutting down. The Soviet Union's run of bad luck continued as Mars 4, 5, 6 and 7 all resulted in failure. Although the Soviets tried their hardest to capture the first unblurred pictures of the Martian surface, the honor eventually went to the United States. Viking 1 landed on Mars in July 1976, just ahead of Viking 2, which landed two months later. Although they were only mandated for ninety days, Viking 1 ended up mapping Mars for six years and Viking 2 for four years. Although their search for life proved inconclusive they gave the world a remarkable insight into the Martian surface.

ABOVE: Martian plain: the Elysium Planitia region of Mars, just north of the Martian equator, is flat and covered in dust, with two impact craters. It is thought that this area, a few tens of kilometers across, is a dust-covered frozen sea. Mars is a cold desert world, with an atmosphere of carbon dioxide and no liquid water.

PATHFINDER MISSIONS

ABOVE: A robotic Sojourner rover vehicle on the surface of Mars sampling the large rock known as "Yogi." The rover was controlled by an operator on Earth. The vehicle weighed 9kg, was 63cm long and 48cm wide, and was powered by a solar panel which allowed for a few hours movement per day.

PHYSICAL DATA

ORBITAL PERIOD (LENGTH OF YEAR) = 687 Earth days

ROTATIONAL PERIOD (LENGTH OF DAY) = 24 hours 38 minutes

After the Viking Missions, it was more than twenty years before America launched another successful probe to Mars. On American Independence Day in 1997 the Mars Pathfinder mission touched down in the Ares Vallis. The arrangement of rocks in this valley seemed to suggest that they had been laid down by liquid water. Unlike the Viking missions, which were stationary, Pathfinder contained a rover, named Sojourner, which gave NASA flexibility to move around the Martian surface and analyze the Martian soil. Pathfinder also discovered an iron core and identified two colors of clouds, one white, comprising water vapor, the other blue, which probably contains carbon dioxide. Shortly after Pathfinder, an orbiter, Mars Global Surveyor, was launched to examine the atmosphere and begin an even more extensive mapping of the surface using its orbital camera to take extremely close-up images of the surface.

SURVEYOR PROGRAM

The Mars Surveyor program of 1998 resulted in disaster for NASA. Two sister probes, the Mars Climate Orbiter and the Mars Polar Lander, were launched separately in December 1998 and January 1999 respectively. The Mars Polar Lander was destined for the south polar cap, and outfitted with an additional probe, Deep Space 2, which was to test for water. While the Mars Polar Lander and Deep Space 2 never reported back to mission control, the Mars Climate Orbiter was lost in the most embarrassing of circumstances. It emerged that while one team at NASA had been working in imperial units, another had been operating a metric system. Consequently, it is thought that the Mars Climate Orbiter burnt up because it entered orbit at far too low an altitude. The debacle meant that NASA ended a century of unprecedented visits to Mars on a low.

ABOVE: Mars Global Surveyor's image of clouds over the surface of Mars. At lower left they are seen covering the peaks of the three large Tharsis Ridge volcanoes. At far left, clouds are also seen around the peak of Olympus Mons while top center the ice cap covering the north pole can be seen.

PHYSICAL DATA

AVERAGE DISTANCE FROM SUN: = 227,936,640km (Earth is 150,000,000km)

AVERAGE SPEED IN ORBITING SUN: = 24 km/sec (Earth = 30km/sec).

DISTANCE FROM THE EARTH (MINIMUM): = 54,510,620 km

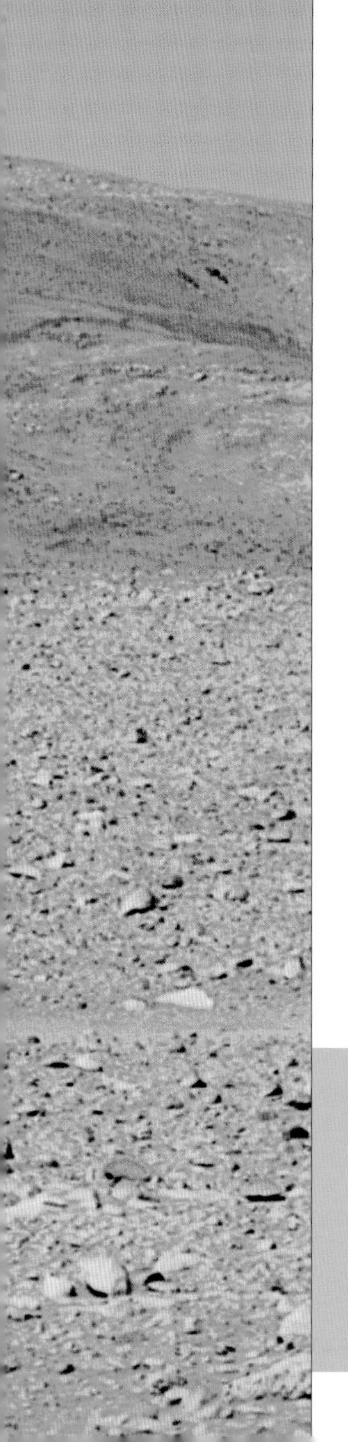

THE TWENTY-FIRST CENTURY

The exploration of Mars should have benefited from the "Mars Rush" of winter 2003-2004.

During this period, Japan's first interplanetary probe, Nozomi, was intended to reach Mars. It was originally destined to enter orbit in 1999 but a series of problems meant that its ETA was revised to winter 2003. However, it was damaged by a solar flare, which caused an electrical malfunction and in December 2003, mission control diverted the probe into space to avoid it crashing into Mars.

Following Japan's Martian foray the European Space Agency, ESA, sent an orbiter, the Mars Express, with a British-led lander mission, Beagle 2, which was to search for signs of life. While the Mars Express continued Odyssey's role of searching for underground water reservoirs, Beagle 2 was lost.

It was NASA, recovering from its woes, who ensured that the Mars Rush was not a complete failure. It successfully sent two more exploration rovers, Spirit and Opportunity, to Mars and improved technology allowed them to cover up to one hundred meters a day. Opportunity gave further evidence of water, as its landing site, Terra Meridiani, appears to have been the shore of an ocean or lake. Both rovers have far exceeded their allotted time of operations and continue to explore and analyze the surface.

LEFT: Columbia Hills on Mars. NASA's Mars Exploration Rover Spirit took the photographs for this composite, true-color image from a distance of 300 meters from the base of the hills.

TIMELINE

1604 - Johannes Kepler calculated an elliptical orbit for Mars.

1666 - Jean-Dominique Cassini (the Italian-born French astronomer) calculated the length of a day on Mars to be 24 hours and 40 minutes.

1671 - Cassini calculated the distance from Earth to Mars.

MEN ON MARS?

On 12 August 2005 NASA launched the Mars Reconnaissance Orbiter. This probe will continue the search for water, past or present, and also act as a telecommunications base for future missions. In 2007 NASA intend to launch their next lander, Phoenix. Unlike Spirit and Opportunity, Phoenix is to be a stationary lander, designed to investigate the northern ice cap. Another NASA rover mission is scheduled for 2009, followed two years later, if all goes well, by a second chance for the European Space Agency to send a roving lander to Mars to search for life.

President George W. Bush announced in 2004 that man would return to the moon by 2020 to create a staging post for a manned mission to Mars. Humans are needed to make faster and more far-reaching discoveries than probes and a manned mission to Mars would proudly denote a key stage in humanity's evolution. Scientists are not sure just how well man would fare on Mars; the atmosphere or the dust might be too challenging for even the best spacesuits. A manned mission would also be extremely costly and many would argue that such vast sums of national revenue might be better spent on healthcare and education on Earth. But it is almost certain that at some point in our future, a human being will stand on Mars.

LEFT: Olympus Mons on Mars. Artwork of Olympus Mons as seen from the north. It is the largest known volcano in the solar system. Its summit rises 27km above the surrounding plains (Mount Everest is 8.8km high), and the base measures 600km across. In front of the volcano is a region of ridges and hills known as the Lycus Sulci (rough area, center).

JUPITER

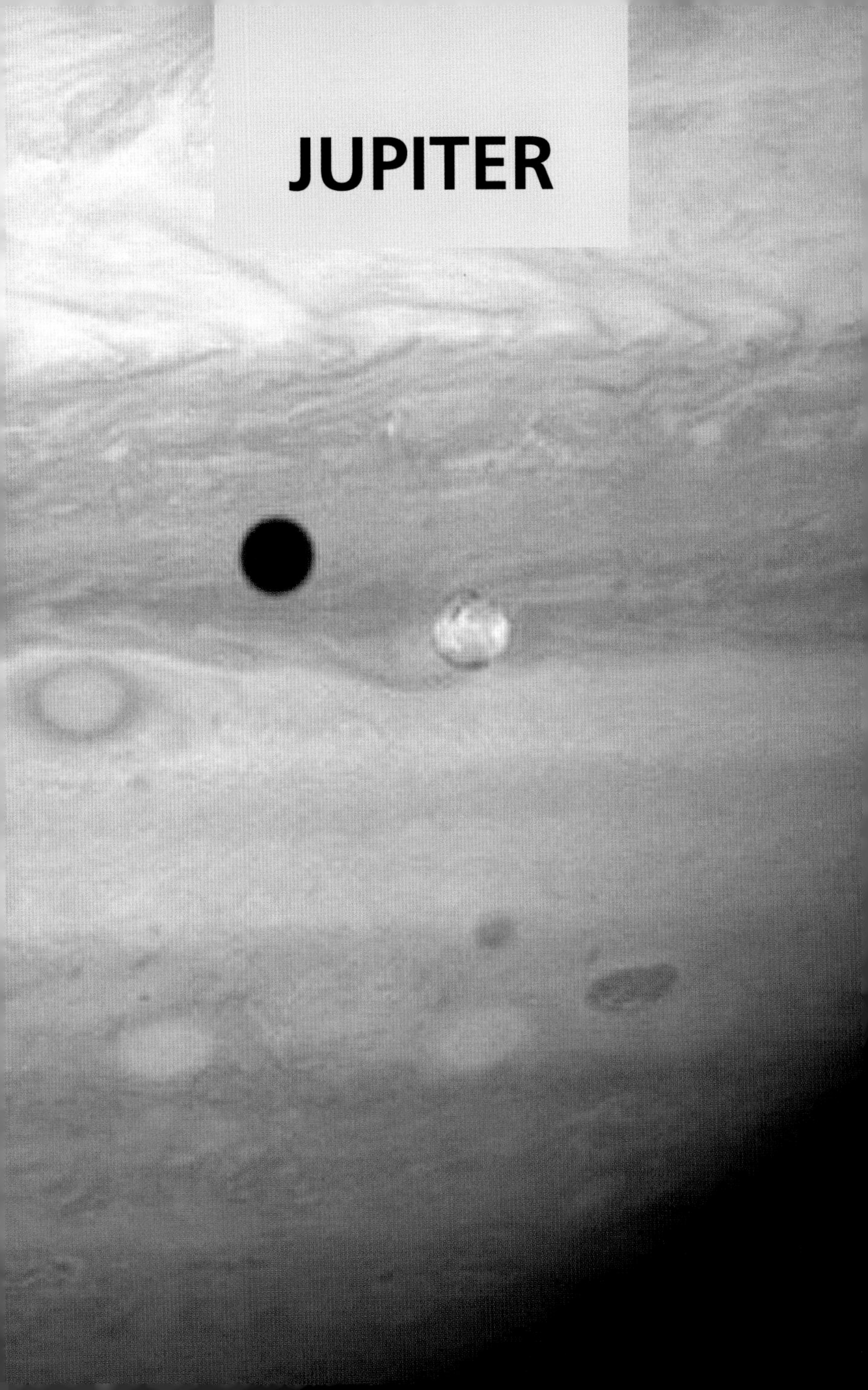

THE GAS GIANT

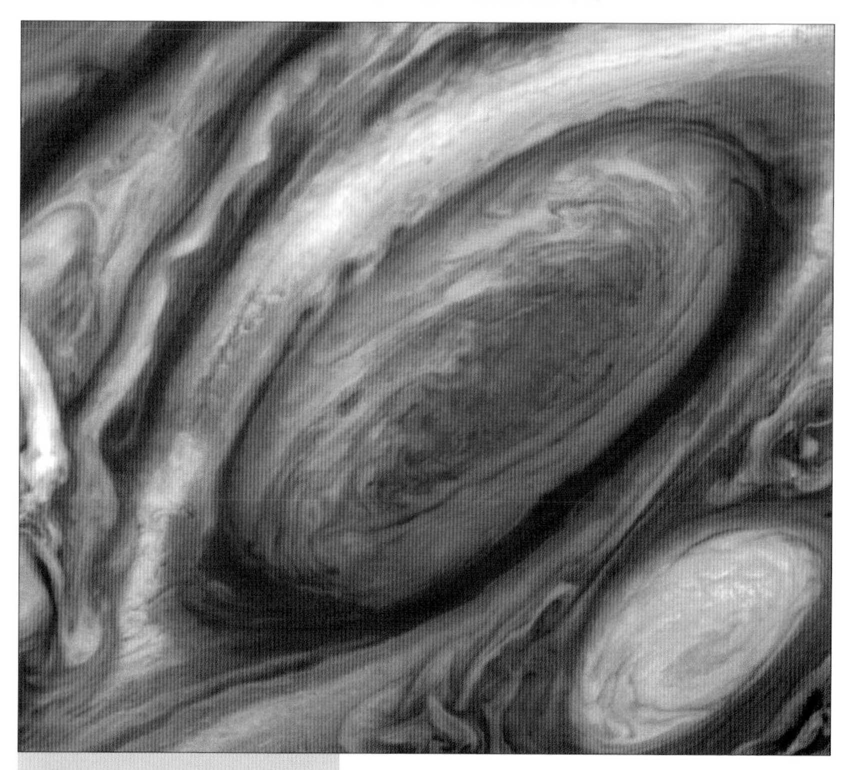

ABOVE: Voyager 2 view of Jupiter's Great Red Spot taken from a distance of 2.6 million kilometers. A long, narrow, white cloud is seen along the Great Red Spot's northern boundary.

PHYSICAL DATA

MASS: = 1.90 x 10^{27}kg; the most massive in our solar system (Earth = 5,97 x 10^{24}kg)

VOLUME: = 1.4 x 10^{15}km^3 (Earth = 1.1 x 10^{12}km^3)

DIAMETER: = 142,984km (Earth = 12,756km)

CIRCUMFERENCE: = 449,197km (Earth = 40,075km)

SURFACE AREA: = 62,179,600km^2 (Earth = 510,072,000km^2)

Jupiter, the fifth planet in our solar system, is so massive that it contains more than twice the mass of all the other eight planets combined, and well over 1000 Earths could fit inside it. As the largest member of the solar system, Jupiter can appear brighter than the brightest star in the sky.

Jupiter is not a perfect sphere. Its equatorial diameter is almost 10km more than its polar diameter. Such flattening of the poles is caused by Jupiter's rapid rotation on its axis – the planet spins at more than 45,000 kilometers per hour. Consequently, it takes the giant just under ten hours to complete one rotation of its axis. The result is that the planet appears to bulge at its equator.

CLOUD DECKS

Jupiter is the first of the four gas giants, the collective name for which, the Jovian planets, is a direct reference to the fact that they are Jupiter-like. Over 80% of Jupiter's atmosphere is composed of hydrogen and much of the remainder is supplied by helium with traces of oxygen, sulfur and nitrogen. The presence of these three substances in the atmosphere leads to three cloud decks. The top layer of clouds is made up of ammonia and is brightly colored. Beneath, in the middle deck, ammonia hydrosulfide crystals give Jupiter its dark red color. The bottom deck appears as bluish cloud, and comprises water which is likely to be frozen, as the temperature is well below zero degrees Celsius.

ABOVE: Voyager 1 picture of Jupiter limb and white ovals.

PREVIOUS PAGES: Southern hemisphere of Jupiter, imaged by NASA's Cassini spacecraft. Jupiter's moon Io and its shadow are also seen.

TIMELINE

364 B.C. - The Chinese astronomer Gan De made an observation of a body which is now believed to be Ganymede – 1,974 years before Galileo.

1664 - British chemist and physicist Robert Hooke discovered the Great Red Spot.

1665 - Cassini measured the rotational rate of Jupiter.

THE GREAT RED SPOT

Jupiter's Great Red Spot is the most notable feature on the planet. It is a giant storm, which flows anti-clockwise across an area in excess of 24,000km in length and 12,000km in width. This giant anticyclone is greater than twice the diameter of the Earth. The storm is known to be at least three hundred and fifty years old. The spot's red color probably comes from the reaction of phosphorus with the sun. It is not only its color which makes the Great Red Spot so pronounced: it is also because the cloud tops of the storm are elevated at least 8km above the surrounding cloud tops. On Earth, hurricanes dissipate when they move over solid land because of the resulting friction, but there is no solid landmass on Jupiter to cause the necessary friction to disperse the storm. Additionally, the storm seems to be being fuelled by heat from Jupiter, rather than the sun. All of this means that there appears to be no reason for the storm to die out. However, the Great Red Spot has been getting smaller during the last one hundred years in which it has been recorded, perhaps indicating that one day it will disappear altogether.

LEFT: True color optical image of Jupiter, taken from a mosaic of shots by the Cassini spacecraft. Powerful jet streams create bands of colored clouds in the planet's atmosphere, dotted with spots marking atmospheric disturbances which can persist for years. The largest such spot is the Great Red Spot (below center). The clouds are made of ammonia, hydrogen sulfide and water.

PHYSICAL DATA

DISTANCE FROM THE EARTH (MINIMUM): = 591,000,000km

DISTANCE FROM THE SUN: = 778,412,020km

AVERAGE SPEED IN ORBITING SUN: = 13 km/sec (Earth = 30 km/sec)

ORBITAL PERIOD (LENGTH OF YEAR): = 4330 Earth-days (11.9 Earth-years)

ROTATIONAL PERIOD (LENGTH OF DAY): = 9 hours 55 minutes

ATMOSPHERE AND INTERIOR

ABOVE: Colored Galileo spacecraft image of bands of cloud in an equatorial region of Jupiter's atmosphere; the colors approximate to natural tones. The colors of the clouds depend upon the exact mix of chemicals within them. The clouds are stretched into fine filaments and curls by violent, turbulent winds in the atmosphere. Observations at these frequencies provide information on the clouds' composition and altitude.

Jupiter's atmosphere is visibly divided into belts and zones. These are caused by rapid winds traveling in opposing directions. In a zone the wind travels from east to west, and the clouds sink. In a belt, the opposite is occurring, the wind is traveling from west to east and the clouds are rising. The zones and belts appear well defined because of the sheer speeds of the winds propelling them. Jupiter does not contain the same layered rocky structure as Earth and the other terrestrial planets. Instead, the interior of Jupiter is mainly composed of hydrogen and helium. Beneath the cloud cover, the pressures on Jupiter are so great that the hydrogen and helium are found in liquid form. As the pressure and heat increase further, to incredible levels, it is believed that the liquid hydrogen might possess the characteristics of a metal, where electrons are able to move about, transferring electric charges between atomic nuclei. In this state the substance is known as liquid metallic hydrogen. Beneath the layer of liquid metallic hydrogen, there is thought to exist a relatively small core containing iron and rocks.

SHOEMAKER-LEVY 9

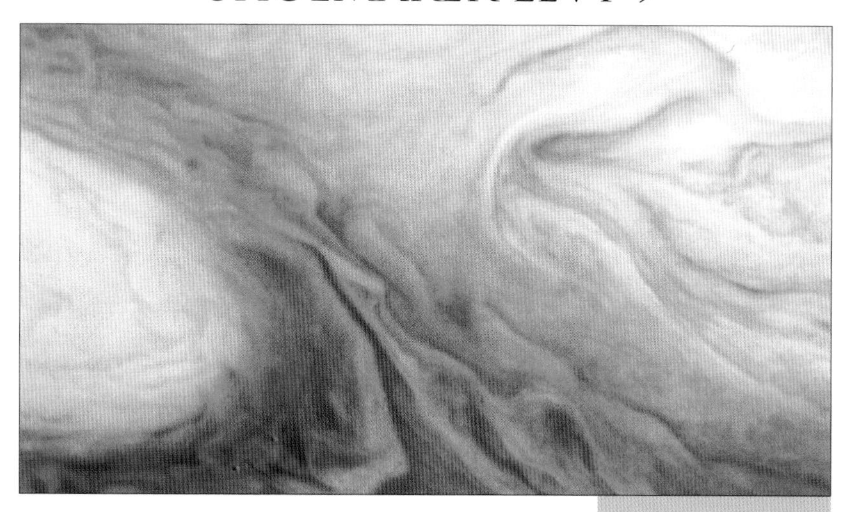

In March 1993, Eugene and Carolyn Shoemaker and David Levy identified their ninth comet, Shoemaker-Levy 9, orbiting Jupiter. It was the first comet observed orbiting a planet rather than the sun. It was soon realized that the fragmented comet was on a collision course with Jupiter. Such a clash between two major objects in the solar system led to a wealth of interest, not just from astronomers, but the general public as well. The impacts took place between 16 and 22 July 1994. Eager astronomers were disappointed; the collision could not be seen from Earth, but by a fortunate coincidence, the Galileo probe was en route to the planet and managed to capture the event for the world to see. The collision resulted in seismic waves across Jupiter and caused huge plumes to rise from the planet's interior, giving scientists a partial idea as to what was below the clouds – an appetizer before Galileo would probe further. For many months after the impacts, large dark spots, more vivid than even the Great Red Spot, could be seen where the comet fragments impacted. The episode highlighted how important Jupiter's gravitational pull has been in protecting the inner planets from many impacts during their history, and that Jupiter deserves some of the credit for sustaining the development of life on Earth.

ABOVE: Voyager 1 image of clouds in the atmosphere of Jupiter, just to the south-east of the giant planet's Great Red Spot. Differences in cloud color may indicate relative heights of the cloud layers.

PHYSICAL DATA

SURFACE GRAVITY: = 20.9m/s (If you weigh 100kg on earth you would weigh 214kg on Jupiter)

DENSITY: = 1.3g/cm^3 (Earth =5.3/cm^3)

ATMOSPHERIC PRESSURE: - varies with depth (the pressure inside Jupiter may be 30 million times greater than the pressure at the Earth's surface)

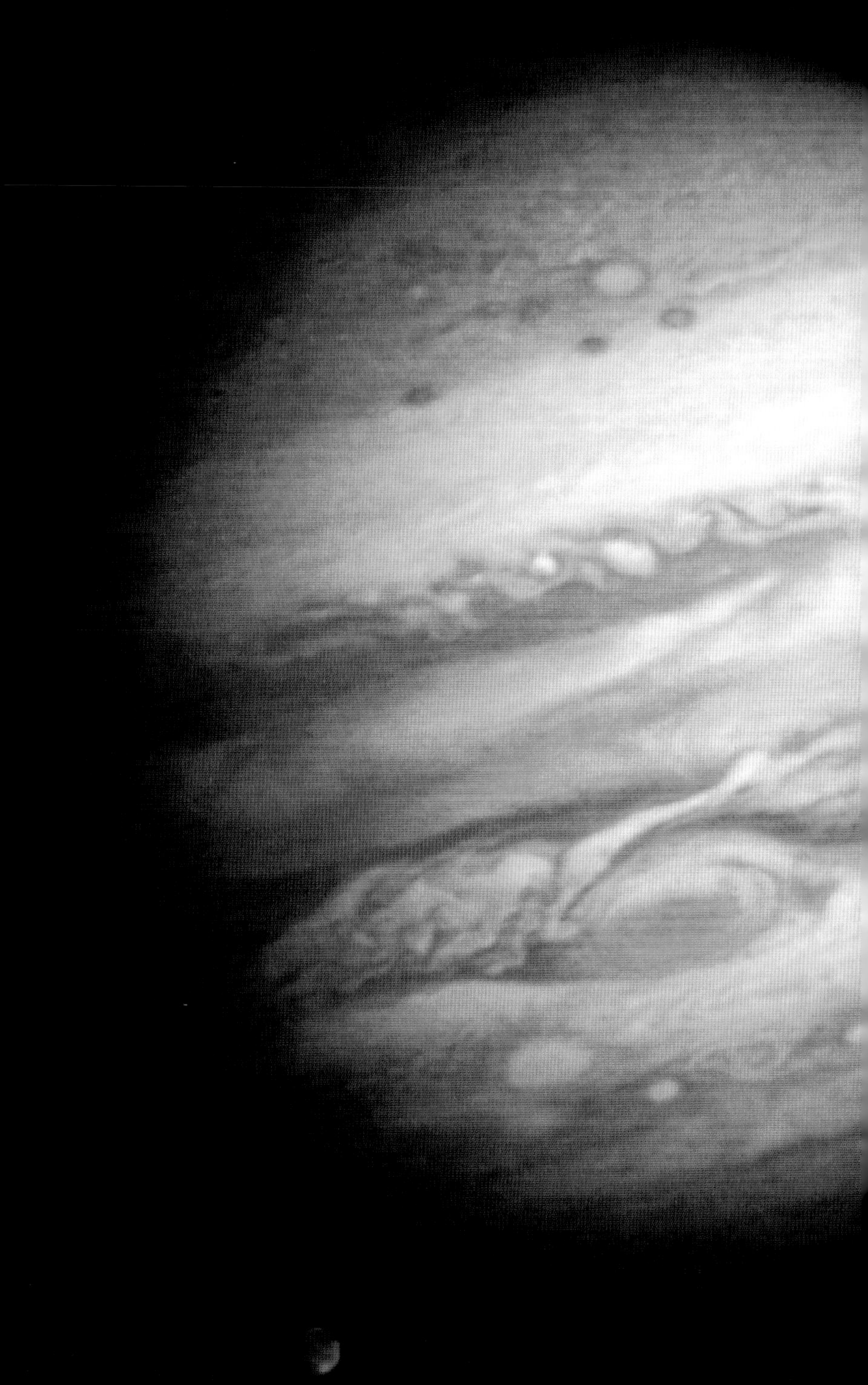

MAGNETOSPHERE

A magnetosphere is the area surrounding a planet in which the planet's magnetic field is stronger than the space around it. Jupiter's is generated by the layer of liquid metallic hydrogen in the planet's interior. Jupiter's magnetosphere is not only the largest magnetosphere of any planet in the solar system, but it is also the largest entity in the solar system, even larger than the Sun – although the Sun's magnetosphere is as large as the solar system itself. If it could be seen by the naked eye on Earth, it would appear larger than the moon in the sky. Jupiter's magnetosphere is so large that its tail reaches beyond the orbit of Saturn, meaning that, for a short time, Saturn can be found inside Jupiter's magnetosphere.

In 1979, the Voyager 1 spacecraft discovered rings encircling Jupiter. Jupiter's rings are much darker than those of its neighbor, Saturn, and this explains why they were not detected from Earth. The inner ring has been named the Halo Ring because it is very faint and cloud-like. The next, called the Main Ring, is over 6000km wide but only 30km thick, giving it a squashed appearance. When the Galileo probe reached Jupiter it also discovered an outer ring, the Gossamer Ring, divided into two sections, surrounding the orbits of the Jovian satellites Thebe and Amalthea.

LEFT: Voyager 1 image of Jupiter from 25 million miles showing Ganymede.

MISSIONS TO JUPITER

1977 MISSION - VOYAGER 2: Voyager 2 achieved a Grand Tour of our solar system, flying by Jupiter, Saturn, Uranus and Neptune. It returned many images of Jupiter and the 4 largest moons

1977 MISSION - VOYAGER 1: Voyager 1 followed Voyager 2 into space within 16 days but arrived at Jupiter 4 months ahead of Voyager 2. Voyager 1 tracked wind speeds and turbulent forms in Jupiter's atmosphere and returned stunning images of the 4 largest moons. Voyager 1 is continuing its journey towards interstellar space and is now further from Earth than any other spacecraft

THE MOONS OF JUPITER

ABOVE: Galileo spacecraft image of two volcanic eruptions on Io, a moon of Jupiter. A 120km high plume from the Pillan Patera volcano appears as a gray patch at Io's lower left edge. The second, from the Prometheus volcano, is visible as a plume (gray) and its shadow at center.

PHYSICAL DATA FOR THE GALILEAN MOONS

GANYMEDE

DENSITY: 1.94g/c^3

ATMOSPHERE: thin oxygen

SURFACE: mountains, valleys, craters

IO

DENSITY: 3.55g/c^3

ATMOSPHERE: sulfur dioxide

SURFACE: volcanic

In1610, Galileo discovered four moons orbiting Jupiter, which were later given the names Europa, Io, Callisto and Ganymede after four of the god Jupiter's many lovers and servants; 282 years later a fifth moon, Amalthea, was discovered. Today, following the Galileo probe's successful mission to the planet, we have knowledge of sixty-three moons. However, the four Galilean moons remain the most interesting.

Io, the closest Galilean satellite to Jupiter, is the most volcanically active body in the solar system. Io's vigorous geological activity is caused by tidal heating by Jupiter, Europa and Ganymede. Callisto is the furthest from Jupiter and is a much cooler world which has lacked sufficient internal heat to remould its surface following impacts from space debris and therefore Callisto's surface is the most cratered in the solar system.

THE GALILEAN MOONS

Both Callisto and Ganymede have a thick ice sheet on their surfaces, but Ganymede is not nearly as uniform as Callisto. There are highlands, but also very old areas that have succumbed to impacts similar to those on Callisto. Ganymede is the largest moon in the solar system; it has its own magnetosphere and is larger than both Mercury and Pluto. Perhaps the most exciting Galilean moon is Europa, because it may be the most promising place in the solar system in which to find the existence of extra-terrestrial life. Visible streaks across the moon's icy surface bear a resemblance to patterns formed by ice in the salt-water seas of the Earth.

ABOVE: Full-disc image of Jupiter's satellite Io.The circular feature at center with a dark spot in the middle is an active volcano, and so are the other features similar to it. Io's volcanic activity appears to be of at least two types: explosive eruptions that hurl material up to 250 km into the satellite's sky; and lava that flows across its surface.

PHYSICAL DATA FOR THE GALILEAN MOONS

EUROPA

DENSITY: 3.01g/c^3

ATMOSPHERE: oxygen

SURFACE: no craters

CALLISTO

DENSITY: 1.86g/c^3

ATMOSPHERE: carbon dioxide

SURFACE: heavily cratered

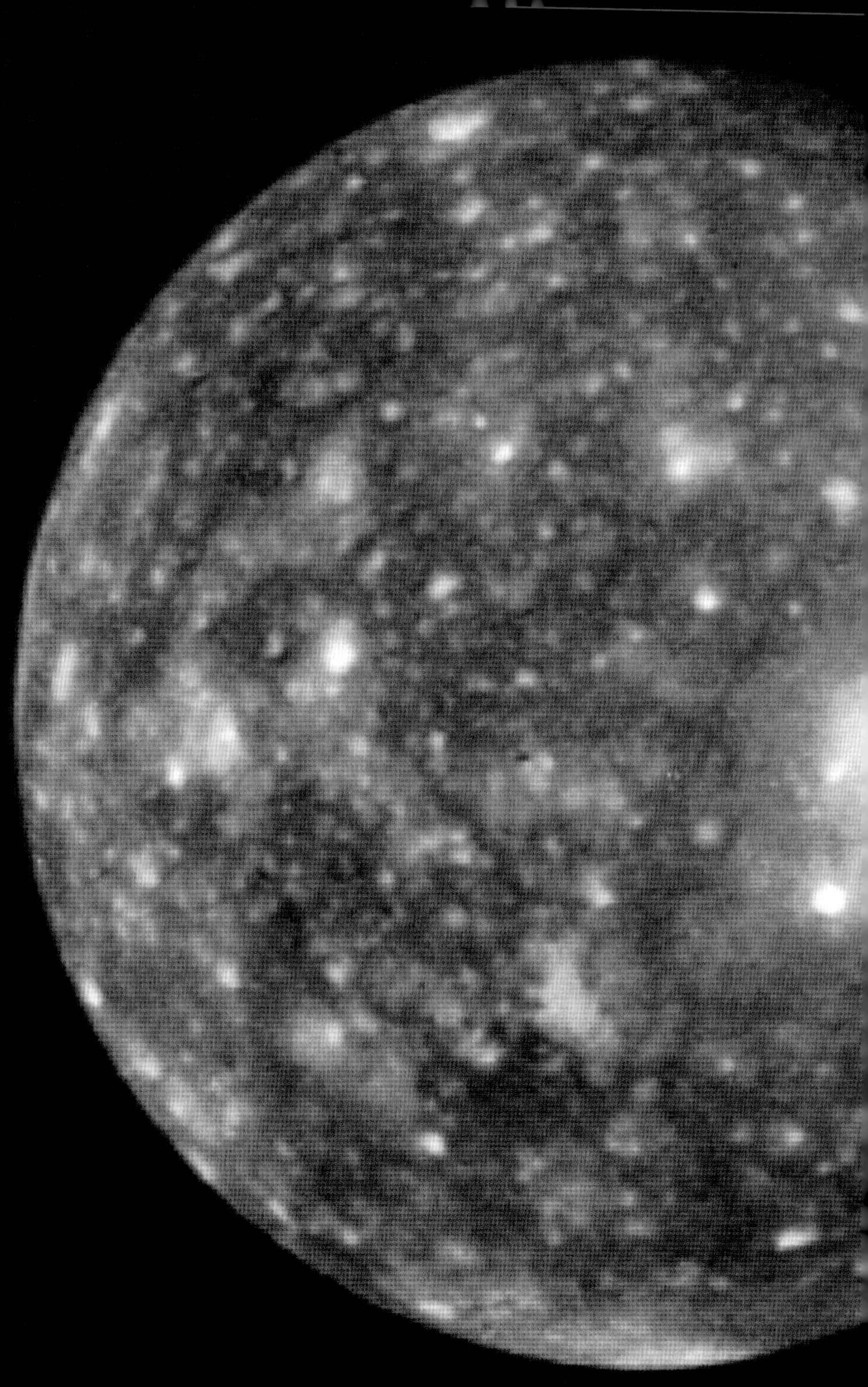

THE NON-GALILEAN MOONS

Jupiter's moons are found in clusters, enabling them to be easily divided into six groups, including the Galileans. Each group bears the name of its largest and most dominant member. The moons within each group are clustered into similar orbital radii and it is thought that each group shares a common origin.

Moons nearer to Jupiter than the Galileans, with orbital radii less than 200,000km, are named the Almathea group. As well as Almathea, this group includes Metis, the closest known satellite to Jupiter, and Adrastea and Thebe. All three were only discovered in 1979 when the Voyagers reached Jupiter.

The next group are the Galileans; the innermost, Io, orbits Jupiter at a mean distance of 420,000km and the outermost, Callisto, at a radius of nearly 2 million km.

The remaining four groups are much further out. Leda, the first moon in the Himalia group is 9 million kilometers beyond Callisto. There are three further groups; the Ananke group containing sixteen moons and the Carme group containing eighteen, each with orbital inclination close to 165. The outermost group is named Pasiphaë; it is not as clustered as the inner groups. Sinope was long thought to be the outermost member of this group, and as such Jupiter's outermost moon, but three more have been discovered even further away.

LEFT: Computer-enhanced photograph of Callisto, the fourth and faintest Galilean satellite of Jupiter. The spacecraft was 2.3 million kilometers from Callisto when the photograph was taken. Callisto is an airless body, where surface temperatures are never higher than -130 degrees Celsius. It is believed that the moon has a large rocky core and a crust that consists of a mixture of water, ice and rocky material. The only surface features evident on Callisto are numerous impact craters formed by meteorites and a number of bright concentric ring systems.

SATURN

THE BEAUTIFUL PLANET

ABOVE: Aurora on Saturn. Aurorae are produced by the interaction of the solar wind with a planet's atmosphere. Charged particles collide with rarefied gases in the atmosphere, causing them to emit light.

PHYSICAL DATA

DISTANCE FROM THE EARTH (MINIMUM): = 1,277,400,000km

AVERAGE DISTANCE FROM SUN: = 1,426,725,400km (Earth = 149,597,000km)

AVERAGE SPEED IN ORBITING SUN: = 10 km/sec (Earth = 30 km/sec)

Saturn is the second-largest planet in the solar system and until 1781, when Uranus was discovered, it was also the outermost known planet. Saturn is the furthest planet visible with the naked eye. When it is isolated from areas packed with stars, Saturn shines strikingly, and is arguably the most impressive sight to be viewed through an amateur telescope, especially if its rings are presented towards the Earth. The planet itself is seen as a yellowish disc with bands and zones of clouds, albeit more subtle than those on neighboring Jupiter. The planet was named Kronos by the Ancient Greeks, and this was changed to Saturn by the Romans in line with their own name for that same god. In Greek mythology, Kronos was the father of Zeus, who was known as Jupiter to the Romans.

ROTATION

It takes Saturn 29.5 Earth-years to complete one orbit of the sun, making its year almost three times longer than a year on Jupiter. Saturn requires just over ten hours to complete one rotation of its axis because the planet is made up almost entirely of liquids and gases which flow around the axis much faster than solid land masses do. This rapid rotation causes Saturn to flatten at the poles and bulge at the equator, like Jupiter – the equatorial diameter is estimated to be 121,000km which reduces to just 108,000km between the poles. Saturn has the lowest density of any planet in the solar system, lower than the density of water. It is regularly remarked that if a big enough ocean could be found, Saturn would float.

ABOVE: Hazy Titan atmosphere showing the night-time side of Titan with a thin sliver of light hitting the visible edge. Above this are layers of haze, reflecting the light from the sun.

PHYSICAL DATA

ORBITAL PERIOD: = 10,756 Earth-days (29.5 Earth-years)

ROTATIONAL PERIOD: = 10 hours 40 mins

SURFACE GRAVITY: = 7.2 m/s^2 (If you weigh 100 kg on Earth you would weigh 74 kg on Saturn)

DENSITY: = 0.7 g/c^3 (about one-eighth as dense as the Earth, and about two-thirds as dense as water)

ATMOSPHERE

It is believed that Saturn's atmosphere is very similar to Jupiter's. It mainly comprises hydrogen with a significant amount – around 5% – of helium, with traces of ammonia and hydrocarbons such as ethane and acetylene. It is thought Saturn's atmosphere contains more sulfur than Jupiter, explaining the planet's yellowish hue. The cloud decks on Saturn are similar to those on Jupiter. There are thought to be three principal decks: a layer of ammonia clouds in the upper reaches of the atmosphere; a layer of ammonia hydrosulfide clouds sandwiched in the middle; and a layer of water clouds. The different cloud strata result from the varying temperatures at which each of the molecules condenses. Within the cloud decks, it appears that temperature increases with depth, indicating that Saturn has an internal source of heat. Saturn's axis is tilted about 27 degrees to the perpendicular of the orbital plane, meaning that Saturn experiences Earth-like seasons. Like Jupiter, Saturn is banded into zones and belts, albeit much less visibly pronounced. The reason for this is rapid winds which cross the planet at speeds of up to 500 meters per second pushing the air in one direction in a belt and in the opposing direction in a zone either side, giving rise to the banded appearance of the planet's disc.

A large white spot was detected in the equatorial regions of Saturn's atmosphere in 1990. The spot was thought to be a storm caused by convection currents from the planet's interior dragging water, ammonia and other molecules above the cloud tops where the temperature is much cooler, causing them to freeze, forming whitish clouds amidst the surrounding yellow. The storm is not a new feature on the planet. It has been recorded twice before during the last 125 years, recurring approximately every 57 years. The fact that this time period is nearly two Saturn years, indicates that the White Spot of Saturn might be a recurring feature.

LEFT: Hubble Space Telescope image of the planet Saturn with its rings and two of its moons visible. The planet is viewed at an angle, while its rings are seen edge-on and thus appear as a thin circular band. These rings comprise a sheet of material (rocks coated with ice) orbiting around the equator of Saturn.

PHYSICAL DATA

DIAMETER: = 120,536km (about 10 times greater than the Earth)

CIRCUMFERENCE: = 378,675km (Earth = 40,075km)

SURFACE AREA: = 43,466,000,000km^2 (Earth = 510,072,000 km^2)

MASS: = 5.69 x 10^{26} kg (Saturn is about 95 times as massive as the Earth)

VOLUME: = 8.3 x 10^{14} km^3 (Saturn's volume is 755 times greater than Earth's)

INTERIOR

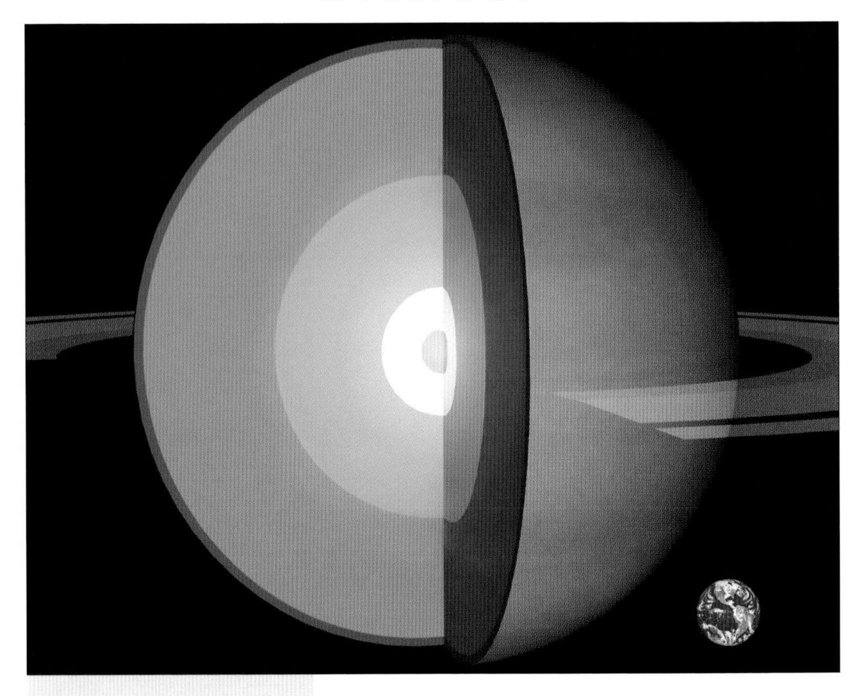

ABOVE: Internal structure of Saturn, cutaway artwork. The Earth is at lower right at the same scale. Saturn is a gas giant.

PHYSICAL STRUCTURE:

Core is solid rock and ice

Layer of liquid metallic hydrogen

Layer of liquid hydrogen

Gaseous atmosphere

The ring system is made up of millions of small chunks of ice and rock

The interior of Saturn is expected to be very similar to that of Jupiter. As atmospheric hydrogen is placed under increasing pressure with depth, it turns into a liquid, where electrons move freely about different nuclei, transferring an electric current. As the layer of electrically charged liquid metallic hydrogen is churned about as a result of Saturn's rapid rotation, a magnetosphere is generated. Saturn's magnetosphere is the second-largest in the solar system, smaller than only Jupiter's, indicating that the layer of liquid metallic hydrogen on Saturn is much thinner than on Jupiter. This liquid metallic hydrogen envelops a rocky core, which is probably larger than Jupiter's because the pressure at the center of Saturn, although intense, is not quite as severe as at the heart of Jupiter.

THE RINGS

One of the greatest fascinations of the solar system is the rings of Saturn. They are usually visible with an amateur telescope, provided the planet is at the correct angle. The rings are located in the plane of Saturn's equator, and like the equator they are inclined at 27 degrees to the orbital plane, which means that for a short period the rings are facing an Earth-based observer edge-on when they are too thin to be seen from Earth. The first person to see the rings of Saturn was Galileo when he was the first observer to look at the planet through a telescope in 1610. He had no idea what he was looking at. Galileo named the rings the "ears" of Saturn, explaining that they might be huge mountain ranges extending high above the planet on both sides, or that they might be two other smaller planets. It was not until 1655, several decades after Galileo's first sighting of the rings, that a Dutch astronomer, Christiaan Huygens, discovered that these "ears" were in fact rings.

ABOVE: Saturn's rings. True color Cassini spacecraft image of a section of Saturn's rings. The image mainly shows the B-ring, the outermost of Saturn's inner rings. Saturn's rings are mostly made of particles of water ice, ranging in size from centimeters to a few meters in diameter.

Ring Name	Width of ring
D	8500 km
C	17,500
B	25,500 km
Cassini Division	4700 km
A	14,600 km
F	30 km - 500 km
G	8000 km
E	300,000 km

RING FORMATION

ABOVE: Saturn and two of its moons, Tethys (top) and Dione. The shadows of Saturn's three bright rings and of Tethys are cast onto the cloud tops. The large gap in the rings is the Cassini Division, which separates the narrower A-ring from the broad B-ring. The thin gap near the top edge of the A-ring is the Encke Division, beyond which is the faint and narrow C-ring.

TIMELINE

1610 - Galileo was the first person to observe Saturn's rings; he named them the "ears" of Saturn believing them to be either huge mountains or two smaller planets

The rings of Saturn are thought to be the remnants of a moon that strayed too close to the planet and was torn apart by tidal forces. The boundary at which it becomes "too close" for a moon to orbit a planet is called the Roche Limit and was calculated by Edouard Roche in 1849, basing his evidence on Saturn's rings. Another theory is that the rings comprise debris left over from the creation of Saturn, and a third suggestion is that Saturn, or one of its moons, suffered a great impact at some point in history. However, all of Saturn's rings are located within the Roche Limit, indicating that they were most likely created by a moon that wandered too close. Moreover, Roche's explanation accounts for why the small particles in the ring do not gravitationally recombine to create larger particles or a moon. In spite of their impressive appearance, there is actually very little matter in Saturn's rings. If they were to recombine, they would only produce a satellite a mere 100km in diameter.

SHEPHERD MOONS

Scientists were keen to understand why the ring system remained in place, instead of being pulled towards Saturn or scattering into space. The chief explanation was found in a series of moons, called shepherd moons, which were orbiting within the ring system, using their gravity to maintain the rings' distinct shapes. This theory was proved in 1980 when two Saturnian shepherd moons named Prometheus and Pandora were discovered. Prometheus guided the inside of F-Ring, while Pandora steered the outside of the same, narrow, ribbon-shaped F-ring. In the same year, another moon, named Atlas, was found to be shepherding the outside of the neighboring A-ring. In 1990, revisiting Voyager 2 photographs, Mark Showalter discovered a moon, Pan, shepherding in the Encke Gap.

ABOVE: Saturn (upper right) in the night sky. The faint blue object (center left) is the Crab nebula (M1, NGC 1952), a supernova remnant. Saturn is one of the five brightest planets, and can be seen with the naked eye. It shines by reflected sunlight.

TIMELINE

1655 - Christiaan Huygens suggested that Saturn was surrounded by a solid ring.

1659 - Huygens discovered that Saturn's rings were separate from the planet.

1660 - Jean Chapelain suggested that Saturn's rings may be made up of a large number of very small satellites.

1675 - Jean Dominique Cassini first notes a division in the rings.

1837 - Johann Encke discovers a second division in the rings.

1883 - British astronomer Andrew Ainslie Common took the first photographs of Saturn's rings.

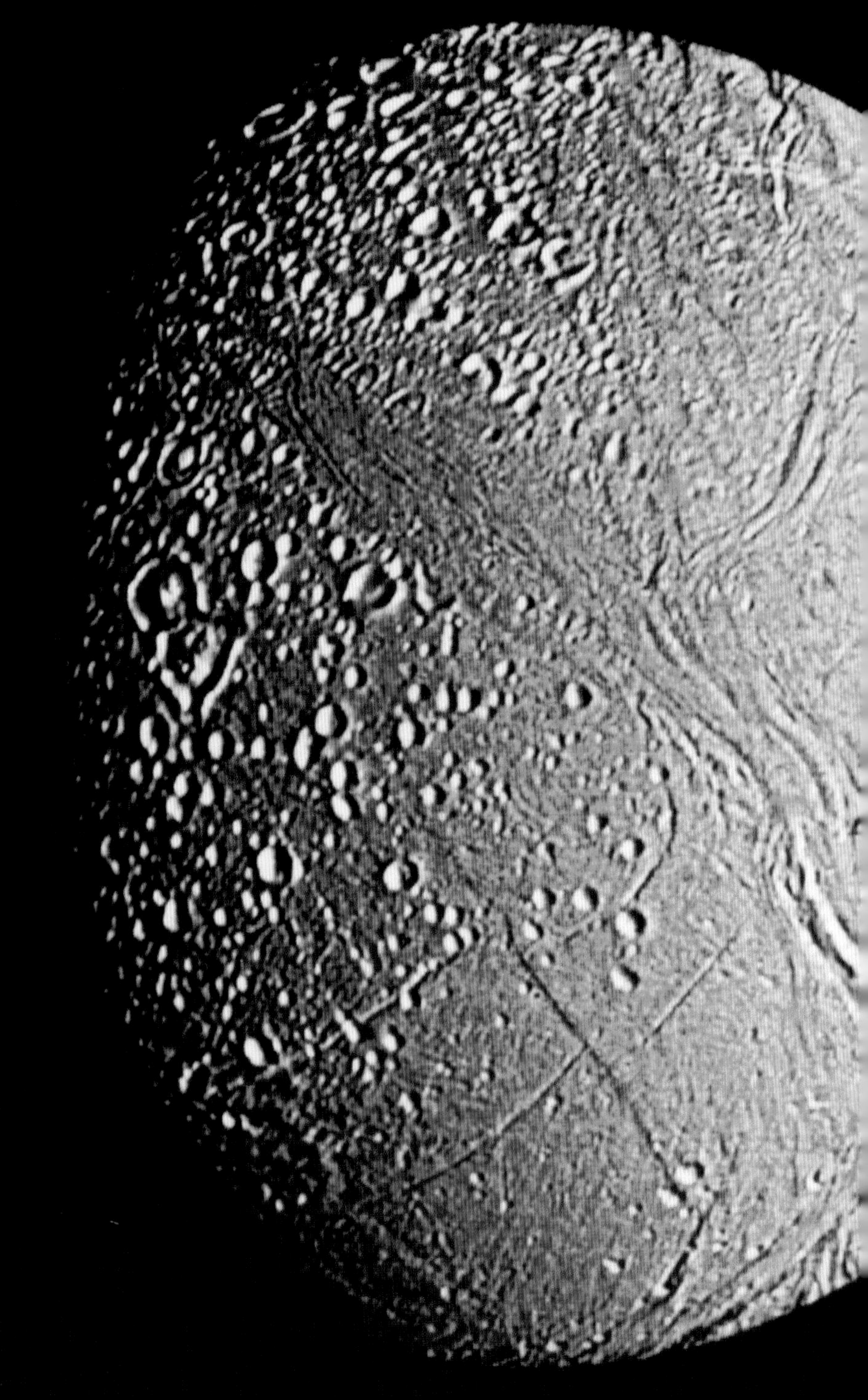

THE SATELLITES

Saturn comes second only to Jupiter in the number of natural satellites it possesses. In 1997 only 18 moons were known about. However more than twice that number have now been identified. The total number of moons currently stands at 48, with more awaiting confirmation. The largest of them, Titan, stands out amongst the planet's satellites – it is much larger and has a denser atmosphere. Saturn has six medium-sized moons: Mimas, Enceladus, Tethys, Dione, Rhea and Iapetus. These moons are all tidally locked with Saturn, so that like Earth's moon, they only ever present one face to the planet. Iapetus, Rhea, Dione and Tethys were discovered by Cassini between 1671 and 1684. Tethys has a density similar to water, indicating that it comprises mainly water ice. The orbits of all the significant Saturnian moons are in the equatorial plane, except Iapetus, the outermost of the medium-sized moons. This means that while from Iapetus the view of Saturn's rings would be exceptional, the perspective from the other moons is less spectacular because the rings are perpetually viewed edge-on.

LEFT: Voyager 2 photograph of Enceladus, a satellite of the planet Saturn. It is 500km across and shows areas that are cratered and others that are criss-crossed with the grooves of canyon-type features.

PHYSICAL DATA

TETHYS: MEAN DISTANCE FROM SATURN: = 294,660km

DIAMETER: = 1,060km

MASS: = 6.27×10^{20}kg

DIONE: MEAN DISTANCE FROM SATURN: = 377,400km

DIAMETER: = 1,120km

MASS: = 1.1×10^{21}kg

RHEA: MEAN DISTANCE FROM SATURN: = 527,040km

DIAMETER: = 1,528km

MASS: = 2.31×10^{21}kg

IAPETUS: MEAN DISTANCE FROM SATURN: = 3,561,300km

DIAMETER: = 1,560km

MASS: = 1.6×10^{21}kg

EARTH'S TRUE SISTER ?

Titan is much larger than the neighboring Saturnian moons. It is larger than Mercury and Pluto and the second-largest moon in the solar system after Jupiter's Ganymede. It was discovered in 1655 by the Dutch astronomer Christiaan Huygens, the first moon to be detected following Galileo's revolutionary discovery of the four moons orbiting Jupiter. Titan is the only body in the solar system to contain a nitrogen-rich atmosphere like the Earth's, although it is much more dense. The atmosphere has prevented observers from seeing what lies beneath on Titan's surface.

After decades of wondering what lay beneath, the European Space Agency's Huygens lander, which piggybacked on the Cassini mission to Saturn, touched down on the surface of Titan early in 2005. Detailed analysis of the data gathered from the probe is yet to be published, but the immediate indications suggest the moon has many Earth-like qualities, such as changing weather patterns and volcanic activity. Gullies and river beds indicate the presence of methane rain (Titan is so cold that any water present would be frozen). It is speculated that the methane rain feeds methane seas. The Huygens lander has also noted volcanic activity which might be caused by Earth-like plate tectonics. In addition, it has identified organic matter on the surface, materials which are crucial building blocks for life and might help scientists better understand the genesis of life on Earth. As data from the Huygens mission is analyzed, it may emerge that Titan, not Venus, is our sister planet in the solar system.

THE SURFACE OF TITAN 2005

OPPOSITE: Titan's atmosphere seen from the Cassini space probe. The colors represent infrared (red and green) and ultraviolet (blue) radiation emitted by Titan's atmosphere. Infrared radiation is where methane in Titan's atmosphere has absorbed light.

LEFT: Titan's surface as seen by the spaceprobe Huygens on 14 January 2005. The image has been processed to give a suitable indication of the actual color on the surface. The surface is darker than expected and consists of a mixture of water and hydrocarbon ice.

PHYSICAL DATA

TITAN

MEAN DISTANCE FROM SATURN: = 1,221,830km

DIAMETER: = 5,510km

MASS: = 1.35×10^{23}kg

MISSIONS TO SATURN

1973: PIONEER 11
First spacecraft to visit Saturn

1977: VOYAGER 2
Visited the four outer planets - Saturn, Uranus, Neptune and Pluto

1977: VOYAGER 1
Returned stunning pictures of Saturn's rings

1997: CASSINI-HUYGENS
The first spacecraft to orbit Saturn

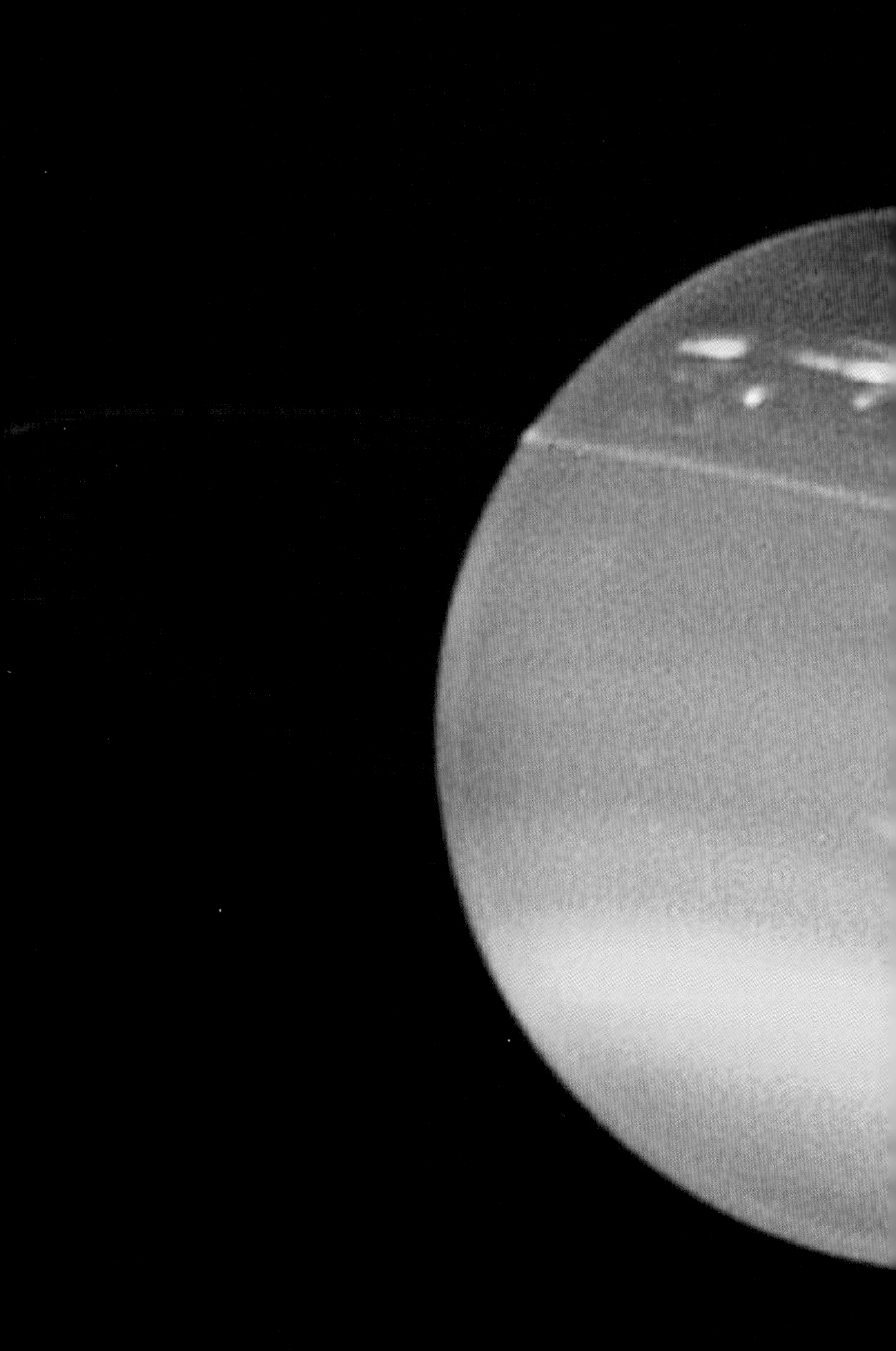

URANUS

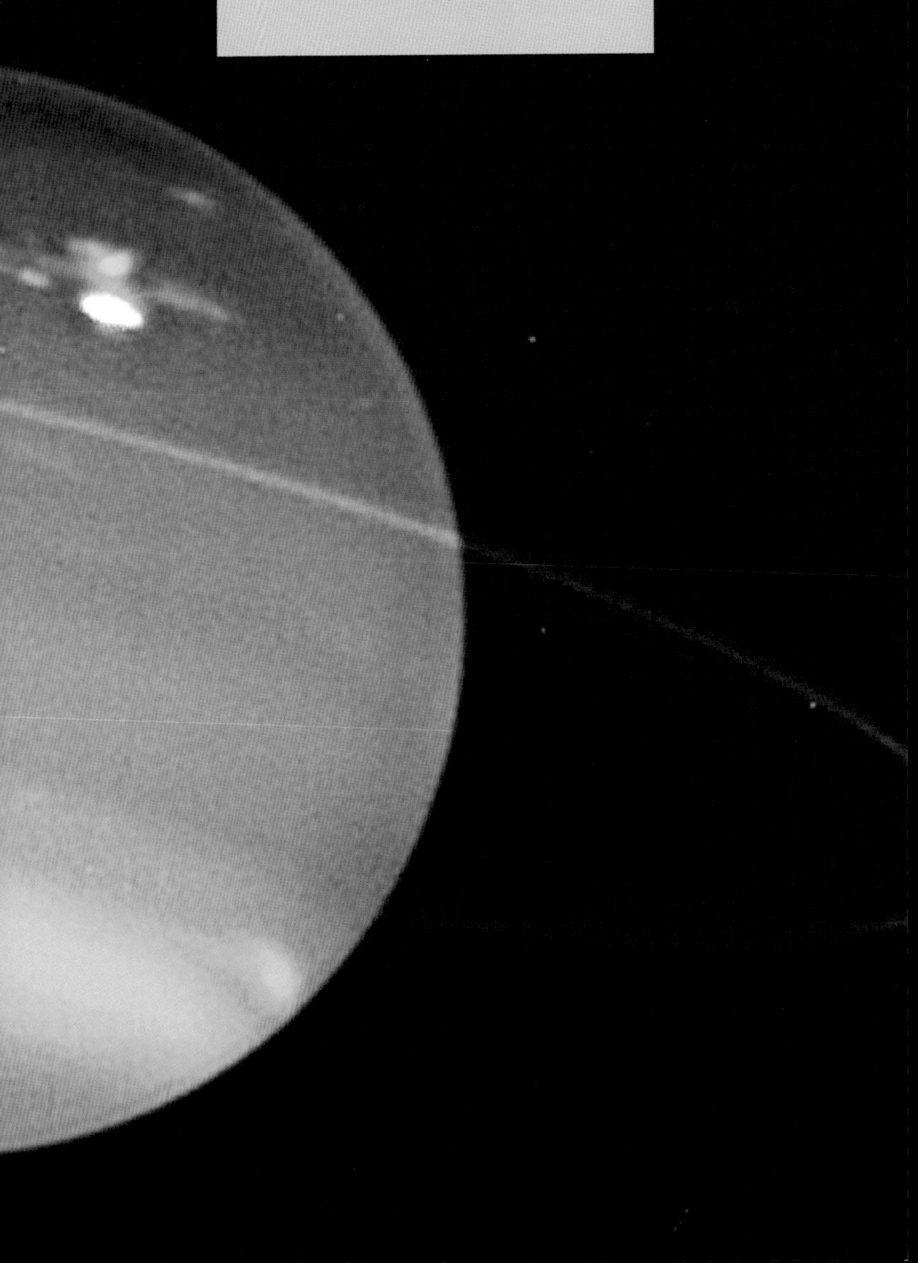

KING GEORGE'S STAR

ABOVE: German-British astronomer Sir Frederick William Herschel who discovered the planet Uranus, resulting in his appointment as private astronomer to King George III of England.

TIMELINE

1781 - Sir William Herschel discovered Uranus.

1787 - Sir William Herschel discovered Uranian moons Titania and Oberon.

1851 - William Lassell discovered Uranian moons Ariel and Umbriel.

On 13th March 1781 a German-born astronomer, William Herschel, discovered what he thought to be a comet. After Herschel had reported his findings it was realized that what he had seen was not a comet at all, but a planet; the first to be revealed by using modern technology, the telescope. As its discoverer, Herschel was given the privilege of naming the planet; he chose Georgium Sidus, George's star, in honor of his king, George III of Britain. Outside Britain, this name was unpopular and the name "Herschel" went into wide use for the planet in the late eighteenth and early nineteenth centuries. However, in order to standardize the new planet's nomenclature with its counterparts in the solar system, the desire for a name from classical mythology arose. In the years following Herschel's death in 1822, the name "Uranus" gained popularity and was finally chosen as the name for the seventh planet.

THE ATMOSPHERE

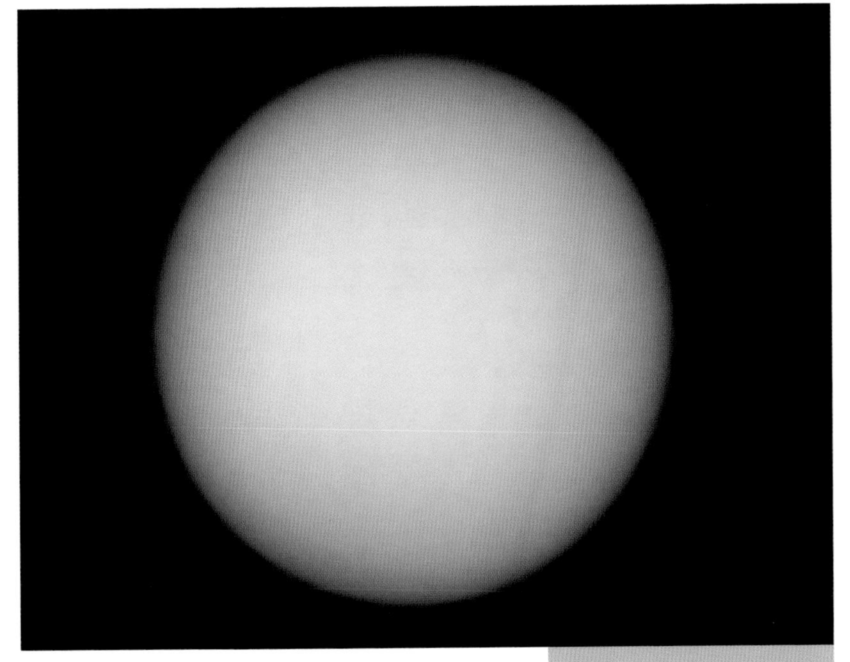

Uranus has perhaps the most featureless exterior of all the known planets in the solar system. Its interior is shrouded by a bluish-greenish blanket, colored by the methane clouds which absorb red light, reflecting blue and green light.Unlike its nearest neighbor, Neptune, Uranus does not appear to have many permanent, noteworthy cloud formations or dark spots. Most of the clouds comprise methane and therefore condense at the same altitude, making the cloud cover relatively uniform across the entire planet. The clouds are moved around the planet by rapid winds which travel at hundreds of kilometers per hour. High-altitude hazes, comprising ethane and other hydrocarbons, have been observed in Uranus' upper atmosphere. Like all the gas giants, the atmosphere comprises mainly hydrogen and helium.

ABOVE: True-color image of the planet Uranus, orbiting at just under 3 billion kilometers, some 20 times the Earth-sun distance. Uranus has an axis of rotation in the same plane as its orbit, so it appears to rotate on its side, unlike the other planets.

PHYSICAL DATA

ORBITAL PERIOD: = 30,687 Earth-days (84 Earth-years)

ROTATIONAL PERIOD: = 17 hours 14 minutess

SURFACE GRAVITY: = 8.4 m/s^2 (If you weigh 100kg on Earth you would weigh 86kg on Uranus)

DENSITY: = 0.7 g/c^3 (about one-eighth as dense as the Earth, and about two-thirds as dense as water)

THE CORE

ABOVE: Color composite image of Uranus. The thin crescent of Uranus, seen here, shows a pale blue-green color. This results from the presence of methane in the planet's atmosphere, which absorbs red wavelengths of light, leaving a predominantly bluish hue.

PHYSICAL DATA

DISTANCE FROM THE EARTH (MINIMUM): = 2,587,000,000km

AVERAGE DISTANCE FROM SUN: = 2,870,972,200km (Earth = 149,597,000 km; Uranus' orbit extends 19 times farther from the Sun than Earth's orbit)

AVERAGE SPEED IN ORBITING SUN: = 7 km/sec (Earth = 30 km/sec)

Temperature and pressure increase with depth on Uranus, with the effect that the methane clouds gradually translate from gaseous states into solid ice. Ice crystals begin developing in the atmosphere, and soon become so abundant that the cloud turns into slushy methane ice, and later solid ice. The solid ice layers comprise water and ammonia as well as methane. These layers flow around the planet at an infinitesimal rate, much as glaciers move around the Earth. At the heart of Uranus is thought to be a rocky core containing silicates and iron. The only probe to have visited Uranus is Voyager 2, when it flew past in 1986 before heading on to Neptune. More probes, destined for Uranus alone, will be required if we are to better understand the structure of the planet's interior and atmosphere.

RINGS

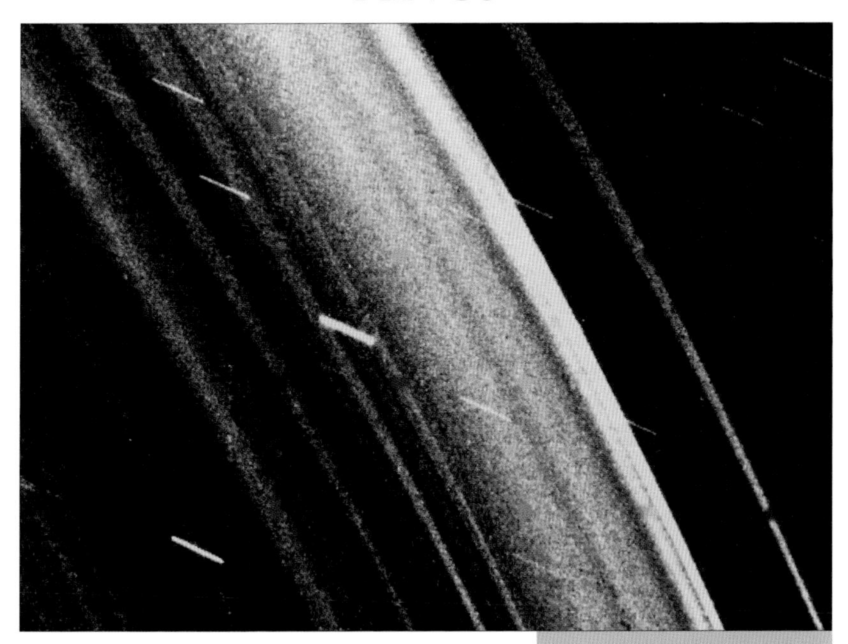

In 1977, many astronomers watched Uranus closely as it passed in front of, or occulted, a star. They wished to calculate the diameter of Uranus by seeing how long the star disappeared as Uranus passed in front of it. As Uranus occulted the star, the light from the star blinked five times each side of the planet. The explanation for such an occurrence was that Uranus was surrounded by a ring system comprising five rings. Later that figure was revised to nine rings, and Voyager 2 confirmed the existence of two more, taking the total number of rings to eleven. The rings are much thinner, darker, fainter and comprise smaller material than the rings of Saturn, which explains why they were not detected before 1977. Cordelia, named after King Lear's youngest daughter, is located within Uranus' ring system and acts as a shepherd for the Epsilon ring, in the same way that several of Saturn's moons shepherd its rings. Cordelia lies within the Roche Limit, and is thought to be close to destruction.

ABOVE: Voyager 2 image of the Uranian ring system. The picture reveals a continuous distribution of small particles throughout the ring system. It also shows all the previously known rings.

RING NAME AND DISTANCE FROM URANUS

1896 U2R	38,000km
6	41,837km
5	42,235km
4	42,572km
Alpha	44,718km
Beta	45,661km
Eta	47,176km
Gamma	47,627km
Delta	48,300km
Lambda	50,024km
Epsilon	51,149km

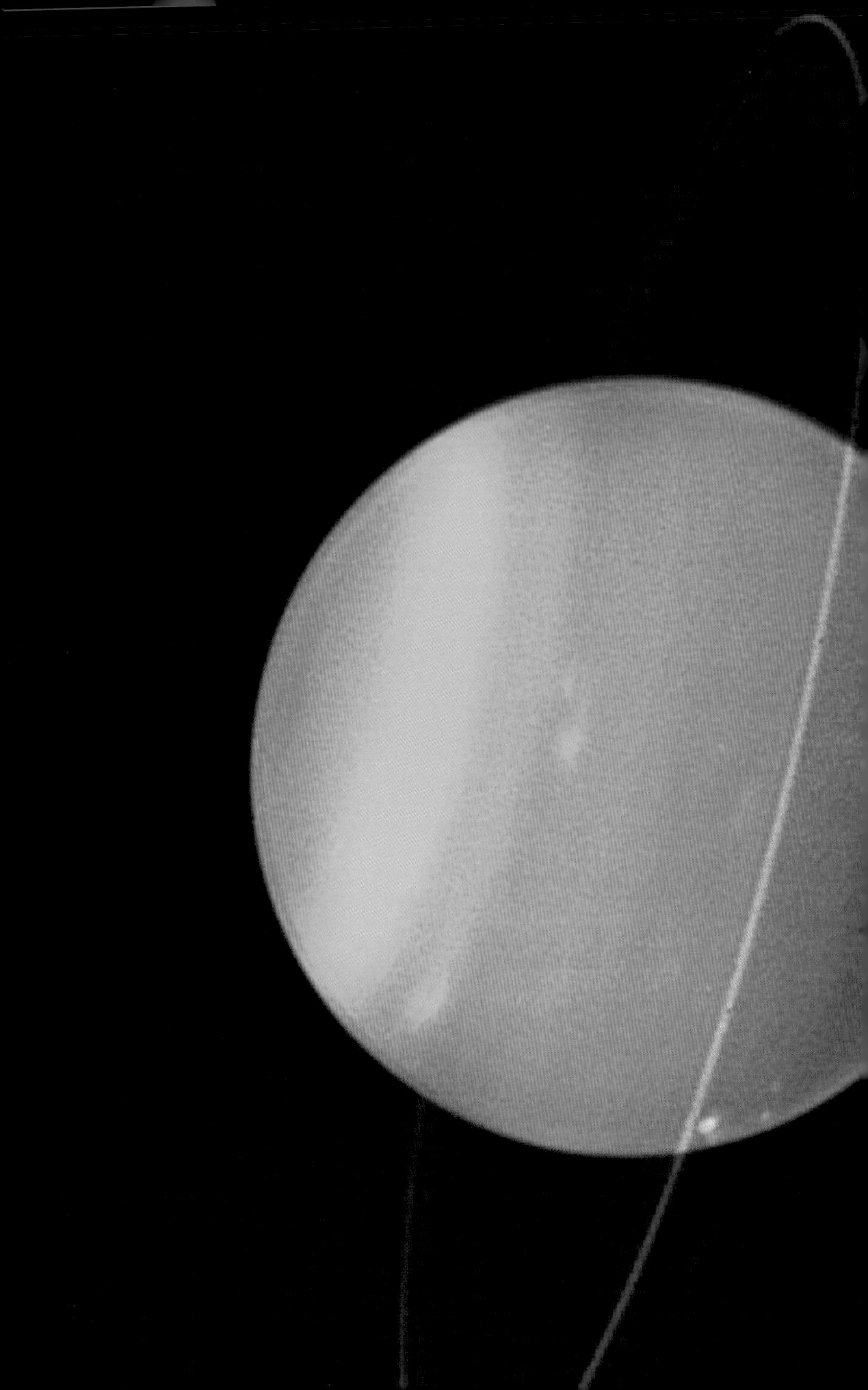

TILTED AXIS

Uranus' most noteworthy feature is the highly eccentric tilt of its axis. The planet's axis is tilted over 90 degrees from the perpendicular, probably because Uranus was impacted by a large object at some point in its history. Such a tilt gives rise to some interesting seasons. The four seasons on Uranus each lasts 21 Earth-years, one quarter of the 84 years it takes Uranus to orbit the sun. At the height of summer, the sun shines directly onto the pole facing it. Meanwhile, the other pole experiences a miserable winter, without sunlight for at least 21 years. In 1985 the south pole moved into the sunshine, leaving the north pole entering a long period of darkness. However, the beginning of the twenty-first century heralded the dawn of spring for Uranus' north pole as the sun moved above the equator. Concurrently, the south pole edged closer to its dark winter. The changing of the seasons was accompanied by large storms in the northern hemisphere, as it began to get warmer – by Uranus' standards.

The storms were photographed by the Hubble Telescope, and may be a recurrent feature of changing seasons on the planet, undetected before because previous seasonal changes could not be observed through earlier Earth-based instruments

LEFT: Uranus, infrared image of Uranus. The northern hemisphere (left of rings) is coming out of many decades of darkness. The bright blue spots in the southern hemisphere are clouds above the Uranian atmosphere. Methane in the upper atmosphere absorbs red light, giving the planet its blue-green color.

PHYSICAL DATA

DIAMETER: = 51,118km (about 4 times greater than the Earth)
CIRCUMFERENCE: = 160,592km (Earth = 40,075km)
SURFACE AREA: = 8,115,600,000km^2 (Earth = 510,072,000km^2)
MASS: = 8.7 x10^{25} kg (Uranus is about 14.5 times as massive as the Earth)
VOLUME: = 5.9 x 10^{13}km^3 (Earth = 1.1 x 10^{13} km^3)

LITERARY MOONS

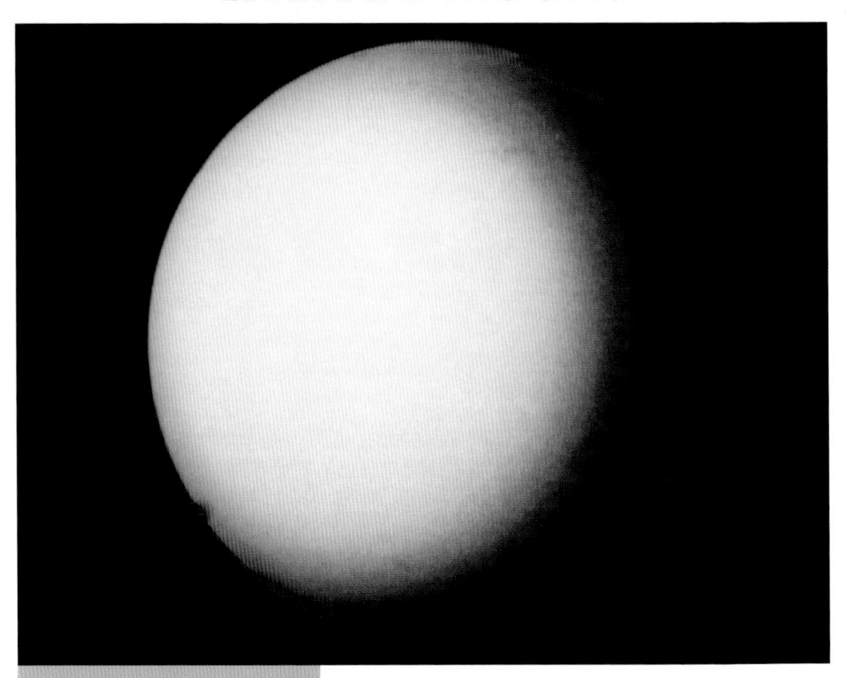

ABOVE: This picture is a composite of images. The pink area centered on the pole is due to the presence of hazes high in the atmosphere that reflect the light before it is absorbed by methane gas in the atmosphere. The bluest regions at mid-latitude represent the most haze-free areas.

TIMELINE

1948 - Gerard Kuiper discovered Uranian moon Miranda

1985 - US astronomer Stephen Synnott discovered Puck from images taken by Voyager 2

1999 - 3 new moons, Setebos, Stephano and Prospero were discovered

2003-6 new satellites were discovered

William Herschel discovered Uranus' first two moons. They were named in 1852 by Herschel's son, John Herschel. He chose Titania and Oberon, after the king and queen of the fairies from William Shakespeare's *A Midsummer Night's Dream*. In 1851, William Lassell discovered two more moons and he requested that John Herschel name them. Herschel chose Ariel and Umbriel, named after characters in Alexander Pope's poem, "The Rape of the Lock." The tradition of naming Uranus' moons after characters from English literature has continued, making the planet's satellites exceptional in a solar system named after figures from classical mythology. Only one more moon, Belinda, was named after a character from "The Rape of the Lock." All the other moons have been named after Shakespearean characters.

ICY CRATERED WORLDS

The largest Uranian moon is Titania, closely followed by Oberon, which has a diameter just 50km smaller. They, like most of Uranus' moons, are icy worlds covered in impact craters. Many of the moons are covered in rift valleys indicating that the rocks may be faulted. Almost one hundred years after Lassell's discovery of Ariel and Umbriel, in 1948 Gerard Kuiper discovered a fifth moon, which he named Miranda, after Prospero's daughter in *The Tempest*. Ten moons were discovered by Voyager 2 as it reached the planet in 1986, and an eleventh moon was discovered over a decade later when a scientist revisited pictures from Voyager 2's mission. In the 1990s and early twenty-first century a further 11 moons were discovered, thanks to improved telescope technology and committed astronomers.

ABOVE: Image of the planet Uranus, showing its ring system and six of its moons. The bright moon at lower right is Ariel. Five other faint moons are seen around the rings. Clockwise from the top, they are: Desdemona, Belinda, Portia, Cressida and Puck.

MISSIONS TO URANUS

Only one spacecraft has observed Uranus at close range - Voyager 2. In 1986 Voyager 2 flew by Uranus at a distance of 107,000km from the center of the planet. Voyager took pictures of 10 new moons in addition to the 5 moons already identified and measured the length of a Uranian day.

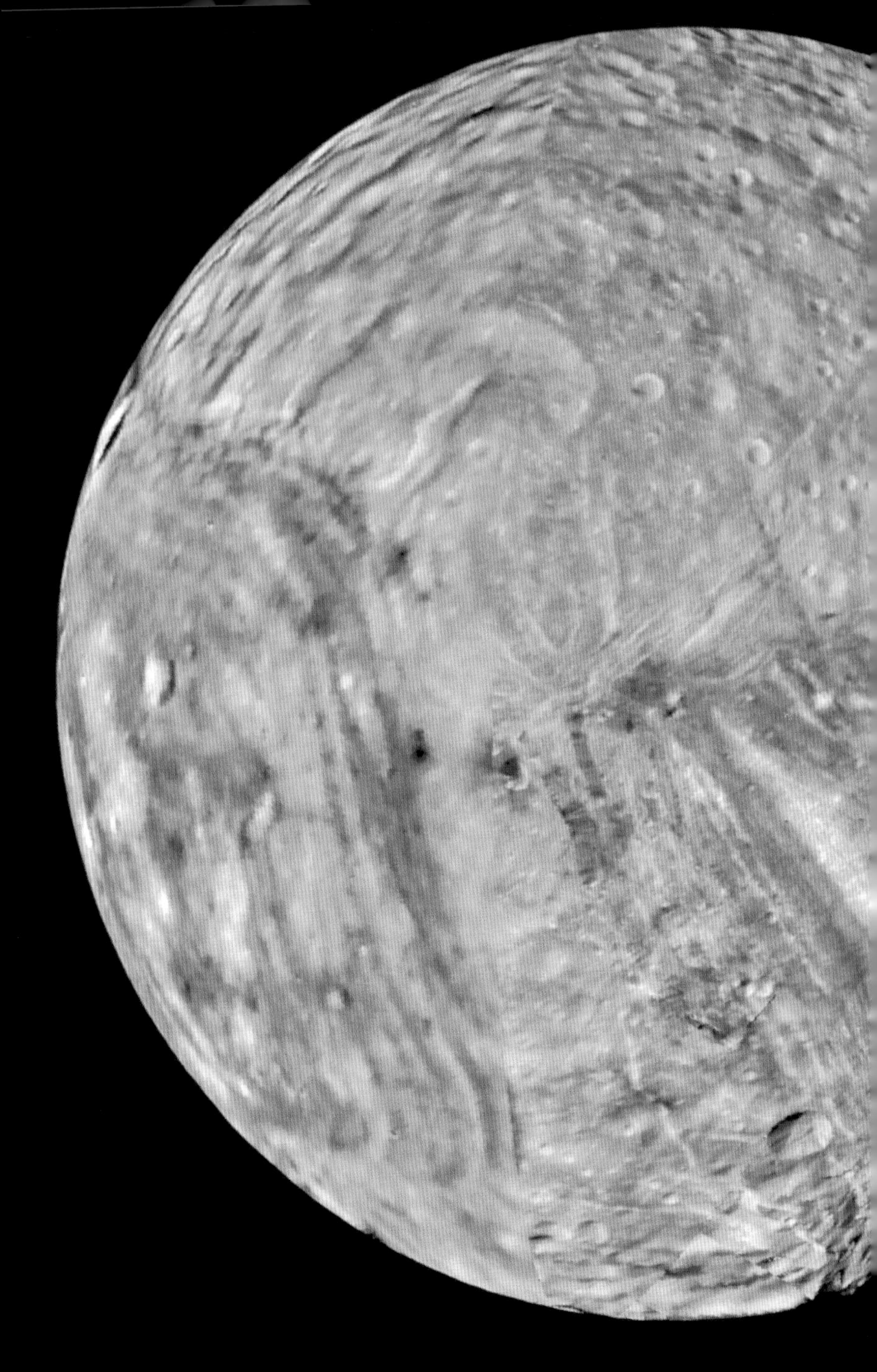

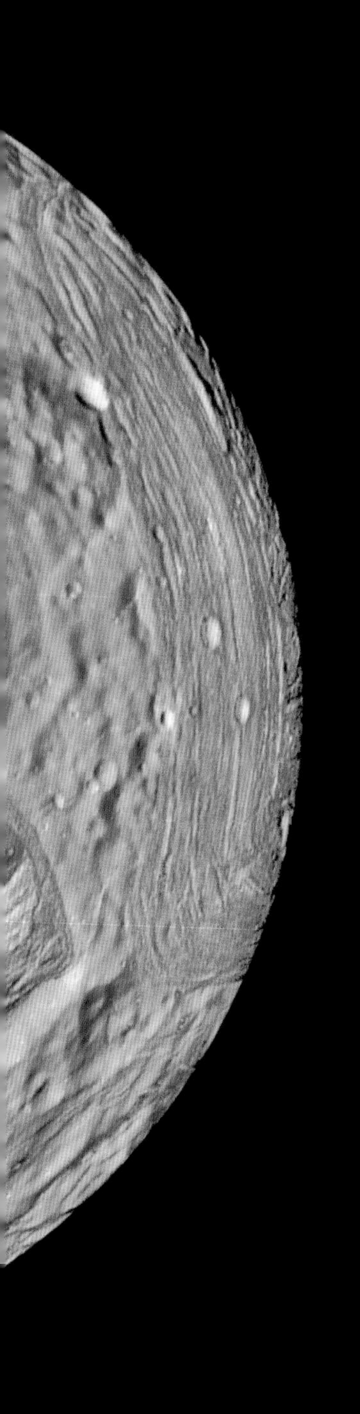

MIRANDA

Of all Uranus' moons, Voyager 2 gave the most insight into Miranda. It would not have been scientists' first choice as it is much smaller than Oberon, Titania, Ariel and Umbriel, but the probe needed to fly close by Miranda in order to reach Neptune, its next destination. However, Miranda turned out to be remarkable. Only the southern hemisphere, which was facing the sun, could be seen by the craft when it passed by in January 1986. Miranda looked like a world which had been completely shattered; it was covered all over in steep ice blocks, cliffs and terraces. One cliff measured 20km high, twice the height of Mount Everest. Initially, scientists believed that Miranda must have been impacted on several occasions by large bodies, which smashed the planet into its current condition. However, more recently, scientists have revised their theories and suggested that many surface features might have been caused by up-welling of warmer ice.

LEFT: Mosaic of images showing the complex surface of Miranda. Only 480km across, Miranda has a V-shaped "chevron" just lower right of center and the "ovoids" at the left limb. There are cliffs near the "chevron" which are twice the height of Mount Everest - about 20km tall.

PHYSICAL DATA

MIRANDA

MEAN DISTANCE FROM URANUS: = 129,872km

DIAMETER: = 480km

MASS: = 6.6×10^{19}kg

DENSITY: = $1.2g/c^{3}$

ORBITAL PERIOD: = 1 day 9 hours 50 minutes

NEPTUNE

DISCOVERED BY MATHEMATICS

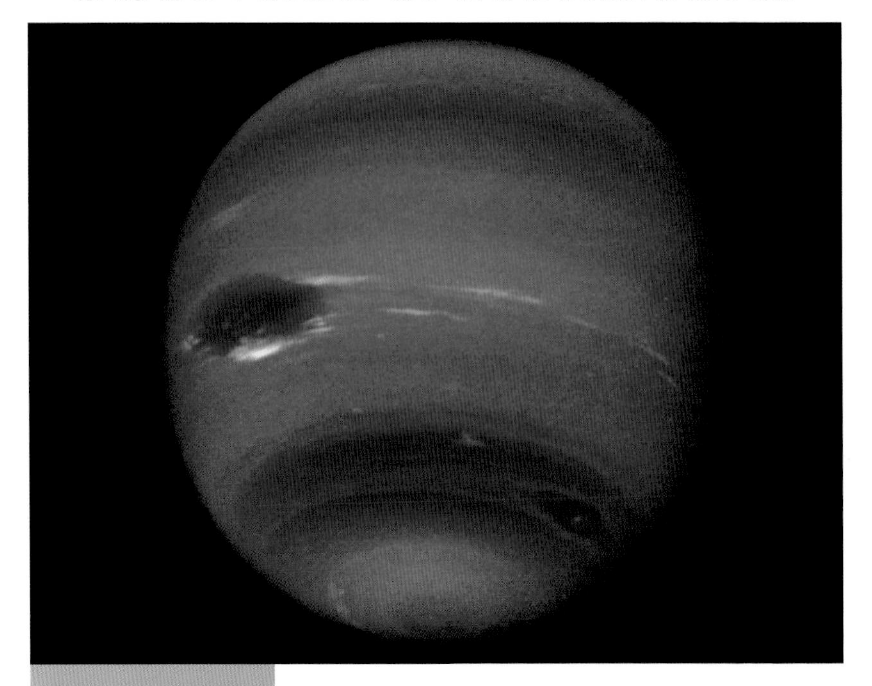

ABOVE: Voyager 2 spacecraft image of Neptune. This virtually true-color image shows two prominent cloud features of Neptune's highly active weather system. At left is the Great Dark Spot, a giant storm system that circuits the planet every 18.3 hours.

TIMELINE

1613 Galileo observed Neptune, but mistook it for a star.

1845 Adams and Leverrier predict the existence of Neptune based on orbital motion of Uranus.

1846 William Lassell discover's Neptune's largest satellite Triton.

Early in the seventeenth century, Galileo recorded an observation of an object in the area of space we now know to be Neptune. However, he took it to be a star. During the nineteenth century, Alexis Bouvard calculated the path that Uranus was expected to take around the sun, but it became apparent that it did not follow such a path, indicating that a planet, beyond Uranus, was pulling it off course. Therefore, the discovery of Neptune was principally the result of mathematics and not observation. Based upon the orbit of Uranus two scientists, John Couch Adams and Urbain Le Verrier, independently calculated where Neptune ought to be. However, neither could generate sufficient interest for observatories to check whether Neptune was in the position both said it would be. Finally, a German astronomer, Johann Gottfried Galle, used Le Verrier's calculations, and sighted the eighth planet of the solar system on 23 September, 1846. Subsequently all three are credited with Neptune's discovery.

THE OUTERMOST PLANET

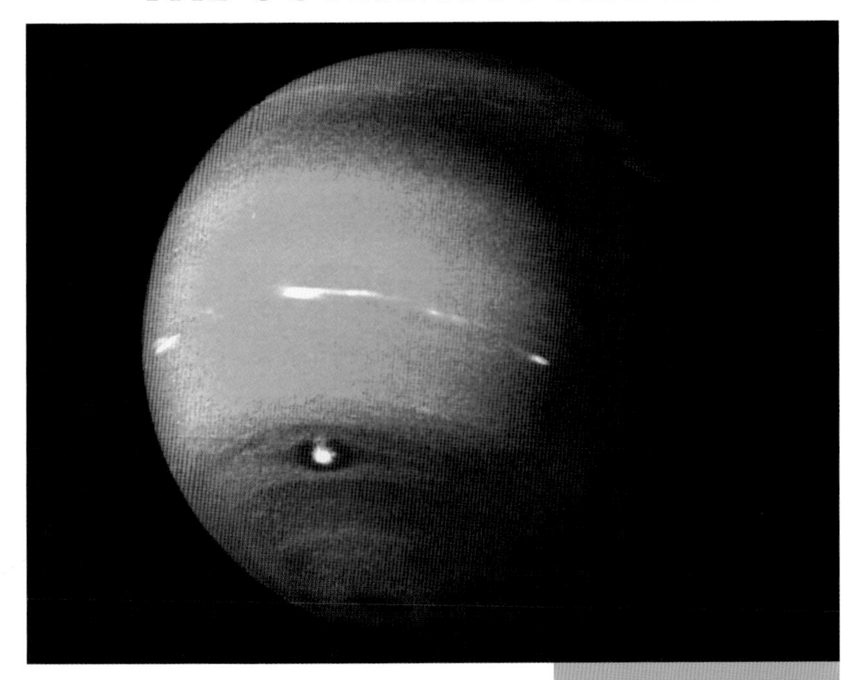

It takes Neptune 164.8 years to orbit the sun, which means that since its discovery, in 1846, it has not yet completed one full orbit of the sun, and is not due to have done so until 2011. During its long orbital period, Pluto's highly eccentric orbit takes it inside the orbit of Neptune, making Neptune the last planet in the solar system for a period of twenty years every 248 years. This occurred very recently. Neptune moved outside Pluto's orbit in 1979 and stayed there until 1999. Therefore, when Voyager 2, the only probe ever to have visited the planet, reached it in 1989 Neptune was the ninth and not the eighth planet in our solar system. The next time Neptune's orbit will go outside that of Pluto's is not until the twenty-third century. Of course, it is in contention as to whether Pluto is a planet at all; if Pluto's planetary status is disregarded, then Neptune becomes the last known planet in the solar system for its entire 165-year orbit.

ABOVE: False-color image of Neptune. It reveals the presence of a ubiquitous haze that covers the planet in a semi-transparent layer. The white areas are high-altitude clouds including one above the center of the Great Dark Spot.

PHYSICAL DATA

ORBITAL PERIOD: = 60,190 Earth-days (165 Earth-years)

ROTATIONAL PERIOD: = 16 hours 7 minutes

SURFACE GRAVITY: = 10.7 m/s^2 (If you weigh 100kg on Earth you would weigh 110kg on Neptune)

DENSITY: = 0.7 g/c^3 (about one-eighth as dense as the Earth, and about two-thirds as dense as water)

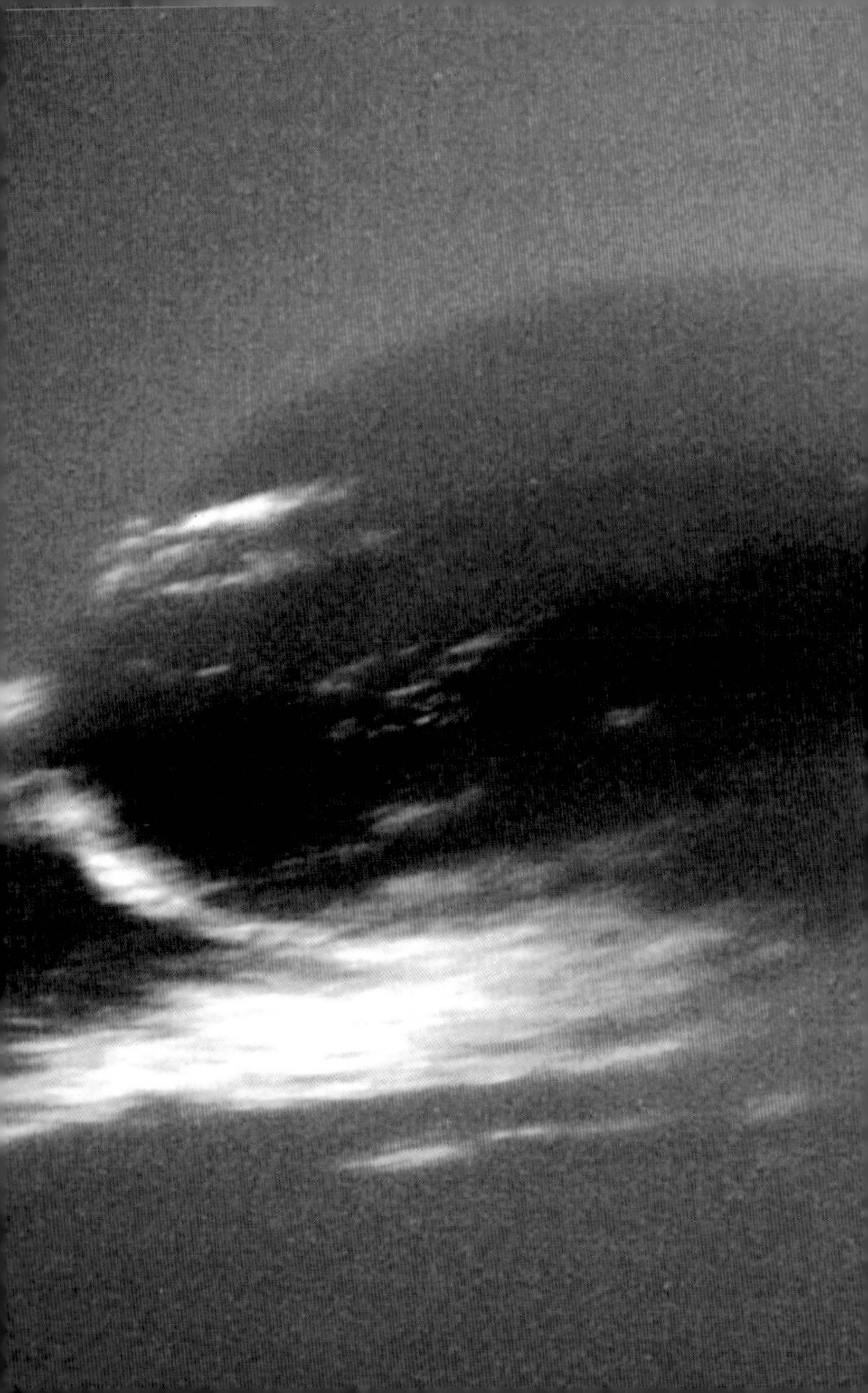

ATMOSPHERE AND INTERIOR

Neptune's atmosphere, like that of all the gas giants, comprises mainly hydrogen and helium gas. However, unlike Jupiter and Saturn, Neptune has noticeable amounts of methane and other hydrocarbons in its atmosphere. Methane clouds, which absorb red light, reflecting green and blue, are thought to be responsible for giving Neptune its blue color. Although initially named a Jovian planet, Neptune has more recently been called a Uranian planet, to indicate Neptune's greater similarity to Uranus, which has an atmosphere similarly comprising 2% methane, rather than Jupiter, which only has trace amounts of the substance.

Neptune is the third most dense planet in the solar system, despite being the fourth largest planet. The gaseous layers of methane, water and ammonia slowly descend to the mantle, first as liquids under great pressure, then slushy ice crystals and finally solid ice. The mantle, which may be enveloped in a layer of liquid hydrogen, surrounds a dense core made of ice and rock.

LEFT: Voyager 2 image of the Great Dark Spot on Neptune photographed from a distance of 2.8 million kilometers. The Spot is a giant storm system as large as Earth. The image shows feathery white clouds that overlie the boundary of the dark blue and light blue regions.

PHYSICAL DATA

DISTANCE FROM THE EARTH (MINIMUM): = 4,310,000,000km

AVERAGE DISTANCE FROM SUN: = 4,498,252,900km (Earth = 149,597,000km)

AVERAGE SPEED IN ORBITING SUN: = 5 km/sec (Earth = 30 km/sec)

METHANE CLOUDS

Neptune has an extremely dynamic weather system, with faster winds than any other planet in the solar system that churn these systems around the planet. Such fast winds may be the result of the temperature difference between the planet's center, warmed by Neptune's internal heat source, and the cooler cloud tops. The winds might also be accelerated by the Neptunian atmosphere, which is so cold that it presents little friction for the winds, which means they can reach the incredible speed of 2000km per hour measured by Voyager 2 in 1989. The winds usually move westward, retrograde to the rotation of the planet, which is from west to east.

Wispy, white clouds, hundreds of kilometers in length, form at high altitudes and can be seen casting their shadow on the bluish layer of gases below. These clouds are similar in appearance to cirrus clouds on Earth, but, unlike cirrus clouds, Neptune's will not be formed from condensed ice water. Instead, they are likely to comprise crystals of frozen methane. One such "cirrus" cloud formation, known as the Scooter, was discovered by Voyager 2 – named because it "scooted" around the planet in just 16 hours.

As well as methane, Neptune's atmosphere comprises other hydrocarbons such as ethane and acetylene. Just as hydrocarbons form smog on Earth, smog or haze has appeared in Neptune's upper atmosphere.

LEFT: Voyager 2 image of white, high-altitude, "cirrus" clouds in the atmosphere above Neptune's southern hemisphere showing the vertical relief in Neptune's atmosphere. The linear cloud formations range in width from 50 to 200km, and their shadow widths range from 30 to 50km. Voyager scientists calculated from this that the cirrus clouds are some 50km higher than the surrounding (blue) cloud deck.

MISSIONS TO NEPTUNE

Voyager 2 is the only spacecraft to have visited Neptune, in 1989, following its fly-by of Jupiter and Saturn. Voyager 2 passed over the north pole of Neptune at a height of 4800km and returned information about the basic characteristics of Neptune and its largest moon, Triton.

THE GREAT DARK SPOT

Neptune's most remarkable feature, a Great Dark Spot in the southern hemisphere, was discovered by Voyager 2 in 1989. It seemed to be similar to the Great Red Spot on Jupiter; rapid winds of over 2000km per hour were discovered in the vicinity of the spot, which meant it was quickly pushed around the planet in just over 18 hours. However, the Spot is now thought to have been a hole in Neptune's atmosphere, similar to the hole in Earth's ozone layer. In 1994, when the Hubble telescope looked for the Great Dark Spot it seemed to have vanished, only to have been replaced with another spot in the northern hemisphere. To differentiate between the two, the original is known as the Great Dark Spot of 1989, while the more recent one is the Great Dark Spot of 1994. Voyager 2 also discovered a similar, but smaller, dark spot in the southern hemisphere. By 1994, it too had disappeared, a testament to the dynamic nature of Neptune's atmosphere.

LEFT: Neptune, taken during August 1989, when Voyager passed within 5000km of Neptune's north pole. This false-color image, obtained using the ultraviolet, violet and green filters of Voyager's wide-angle camera, highlights details of weather systems, revealing clouds located at different altitudes in different colors. The prominent Great Dark Spot (center), a giant storm system the size of Earth, appears surrounded by pink, wispy clouds at high altitudes. Areas of dark, deep-lying cloud appear dark blue in this image.

PHYSICAL DATA

DIAMETER: = 49,528km (the smallest of the 4 giant "gas planets"; Earth = 12,756km)

CIRCUMFERENCE: = 155,597km (Earth = 40,075km)

SURFACE AREA: = 7,640,800,000km (Earth = 510,072,000km^2)

MASS: = 1.02 x 10^{26} kg (Neptune is about 17 times as massive as the Earth)

VOLUME: = 6.3 x 10^{13} km^3 (Earth = 1.1 x 10^{13} km^3 - Neptune could hold nearly 60 Earths)

THE SEASONS

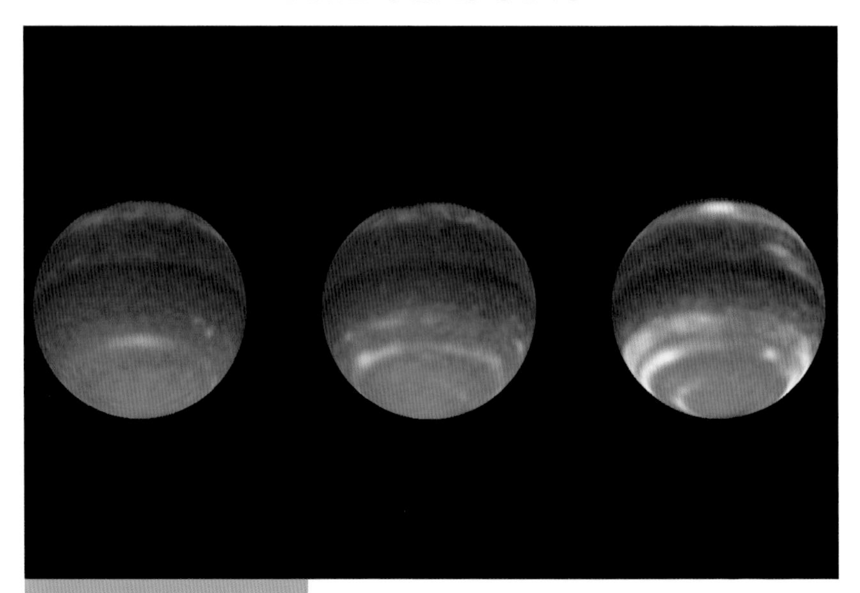

ABOVE: Neptune's seasons. Hubble Space Telescope image of brightening cloud bands in Neptune's southern hemisphere in 1996 (left), 1998 and 2002 (right). This change may be due to the arrival of spring. Neptune has wild weather despite its distance from the sun, and winds can reach 2000km per hour.

PHYSICAL DATA

AVERAGE TEMPERATURE: -2200 C

ATMOSPHERE: hydrogen, helium methane (Neptune's atmosphere has the highest wind speeds in the solar system - up to 2000kph)

Neptune's axis is tilted to 29.5 degrees from the perpendicular, which is similar to Earth, and as such Neptune experiences seasons, as Earth does. Seasons are much less pronounced than on Earth because Neptune is so much further from the sun than the Earth. Each season lasts much longer than on Earth because of the long period of time it takes Neptune to orbit the sun. At the start of the new millennium, growing bands of clouds on Neptune indicated the start of a Neptunian spring, which will last for several decades.

Whatever the season, Neptune's average surface temperature at minus 2200 Celsius is similar to that of Uranus, although it is much further from the sun. This is because Neptune has an internal heat source which compensates for this difference.

TRITON

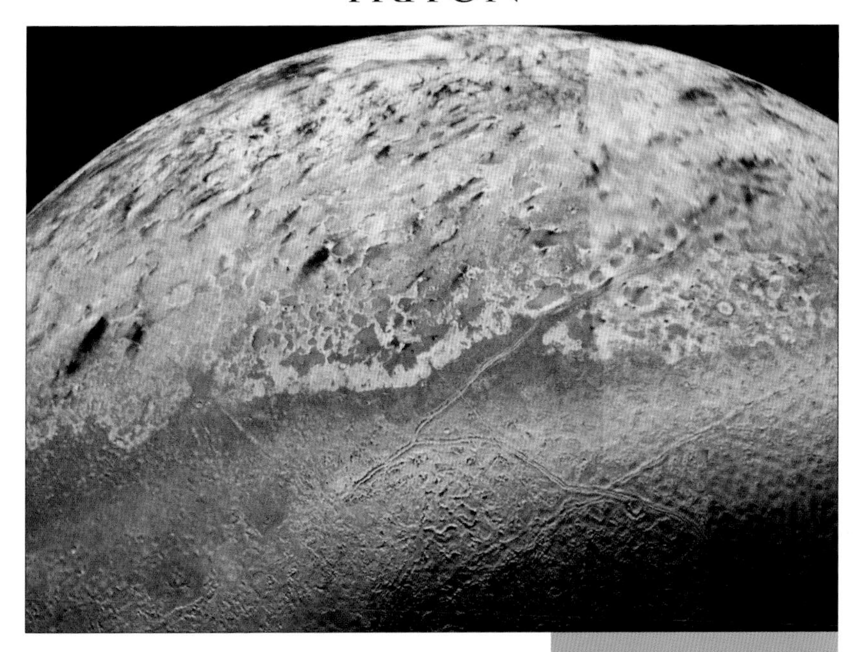

Neptune is known to have thirteen moons. Triton was the first to be observed, just days after the discovery of Neptune by the British astronomer William Lassell. Triton, unlike the other sizeable satellites of solar system planets, orbits Neptune in a retrograde direction, suggesting that the two bodies were not formed together. Before it was captured in Neptune's gravitational field, scientists speculate that Triton would have been much like Pluto, a trans-Neptunian object from the Kuiper Belt. As a relatively large moon which orbits very close to its planet, tidal forces have drawn Triton into an exceptionally uniform, circular orbit. There is an absence of impact craters, indicating that Triton is still geologically active; massive geysers of ice and dust erupt from beneath the surface, forming plumes of over 7km in height. The presence of water, combined with geothermal activity, has led scientists to propose Triton as a possible contender to harbor primitive life.

ABOVE: Mosaic of Voyager 2 images of Triton. The surface shows a great variety of features. The large south polar cap (top) is thought to consist of a slowly evaporating layer of nitrogen ice, deposited during the previous winter. Below this is a region which has been dubbed the "cantaloupe" terrain.

PHYSICAL DATA

TRITON

MEAN DISTANCE FROM NEPTUNE: = 354,760km

DIAMETER: = 2,704km

MASS: = 2.1 x 10^{22}kg

DENSITY: = 2.1g/c^3

ORBITAL PERIOD: = 6 days

PLUTO

THE SMALLEST PLANET

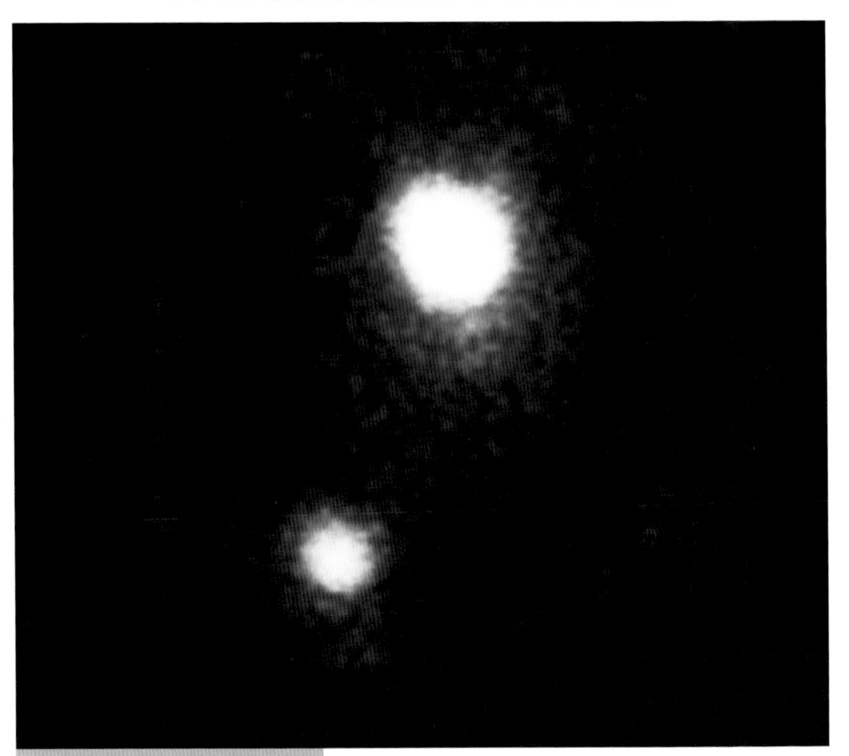

ABOVE: Optical image of the planet Pluto and its large moon Charon taken with the Hubble Space Telescope's Faint Object Camera.

PHYSICAL DATA

ORBITAL PERIOD: = 90,553 Earth-days (248 Earth-years)

ROTATIONAL PERIOD: = 6 days 9 hours 18 minutes

Pluto has an eccentric orbit; its distance to the sun varies between 4.4 and 7.4 billion kilometers

In 1928, an amateur astronomer from Kansas, Clyde Tombaugh, sent detailed observations of planetary movements to the Lowell Observatory in Arizona. Instead of receiving the comments he had hoped for, he was offered a job as a junior astronomer instead. His task was to trawl through countless photographs of the space beyond Neptune in search of an undiscovered ninth planet, named Planet X, a term coined by the founder of the observatory, Percival Lowell. Soon after the discovery of Neptune it was realized that the planet was not massive enough by itself to pull the other gas giants very slightly off their expected courses. Instead the existence of another planet, beyond Neptune, was thought to be responsible.

GOD OF THE UNDERWORLD

Tombaugh carried out many hours of painstaking research, comparing photos of the same area of space at different times to detect the possible movement of a planet against the background stars. On 18 February 1930, he discovered such a movement and thus the ninth planet was discovered. The announcement was made and a competition was launched to name the newest member of the solar system. Venetia Burney, an eleven-year-old schoolgirl from Oxfordshire, England, suggested the name Pluto. In classical mythology, Pluto was the god of the underworld, and this cleverly alluded to the fact that Pluto was so far from the sun. Moreover, the first two letters were the initials of Percival Lowell.

ABOVE: Clyde Tombaugh(1906-1997) with the blink comparator he used to discover Pluto. After 7000 hours of work, he found a tiny object outside the orbit of Neptune, which was named Pluto.

TIMELINE:

1930 - Pluto is discovered by Tombaugh.

1954 - Pluto's period of rotation is determined.

1978 - James Christy discovered Charon.

1992 - Nitrogen and carbon monoxide are discovered on Pluto's surface.

1994 - First Hubble Space Telescope maps of Pluto.

PLANET X?

ABOVE: Pluto on 27 July 1998. Pluto shines at a faint 14.5 magnitude.

PHYSICAL DATA

DIAMETER: = 2300km (the smallest of the planets; it has a diameter about two-thirds of the Earth's moon)

CIRCUMFERENCE: = 7232km (Earth = 40,075km)

SURFACE AREA: = 16,650,000km^2 (Earth = 510,072,000 km^2)

DENSITY: = 2g/cm^3 (Earth = 5.5g/cm^3)

MASS: = 1.3 x 10^{22} kg (due to its low density Pluto has mass about 17% of the Earth's moon)

VOLUME: = 6.4 x 10^9 km^3 (Earth = 1.1 x 10^{13} km^3)

Initially everybody was satisfied that Pluto was Planet X, the ninth planet in the solar system, and several decades later the size of Pluto was calculated when it occulted its moon, Charon. Pluto was found to be much smaller than earlier estimates had suggested. It was not only the smallest planet in the solar system, it was smaller than several moons: Ganymede, Titan, Callisto, Io, Europa, Triton and even Earth's moon. There is still controversy about the ninth planet. Given its exceptionally small size, questions began to emerge as to whether Pluto deserved planetary status at all. Its orbit was discovered to be both eccentric and inclined to the plane in which the other planets orbited, rather more like the orbit of a comet, with a distance of 30 AU when it is closest to the sun, 50 AU at its furthest.

PLUTO'S MOON

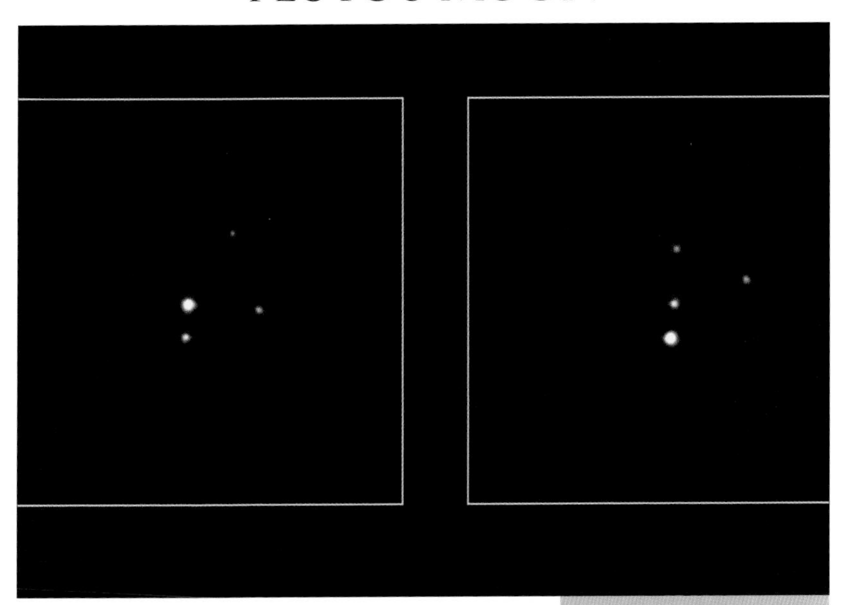

In 1978, American astronomer James Christy discovered that Pluto had a moon when he noticed a small bulge occurring periodically in photographs of the planet. The moon was named Charon, after the boatman who in Greek mythology ferried souls across the River Styx to the underworld, where Pluto was god. Pluto and Charon are perhaps better referred to as a double planet system because both objects are very close to one another and Charon is just under half the size of Pluto, making it the largest moon relative to the size of its parent planet in the solar system. Moreover, not only is Charon tidally locked to Pluto, but Pluto is the only planet tidally locked to its moon. It takes both bodies 6.4 days to rotate on their axis, meaning that one object keeps the same face presented to the other object and vice versa. In 2005, the Hubble Space Telescope captured what is thought to be two new moons orbiting Pluto but the International Astronomical Union cannot confirm and name them until Hubble can take further pictures to prove that the two objects are in fact in the orbit of Pluto.

ABOVE: Images of Pluto (white) and its large moon Charon (pale blue), and two newly discovered moons (gray). Pluto and Charon have such similar masses that they orbit a point in between them, rather than a point within the larger body, as is the case with other moons.

PHYSICAL DATA

CHARON

MEAN DISTANCE FROM PLUTO: = 19,600km

DIAMETER: = 1186km

MASS: = 1.6 x 10^{21}kg

DENSITY: = 2.2g/c^{3}

ORBITAL AND ROTATIONAL PERIOD: = 6 days 9 hours 18 mins (i.e. synchronous with Pluto)

FROZEN NITROGEN

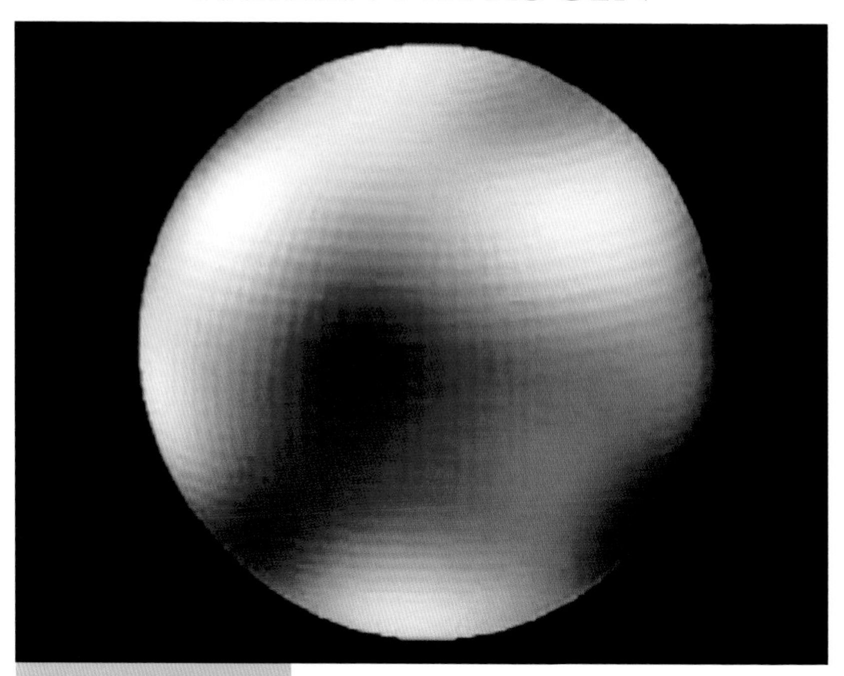

ABOVE: This image was processed to bring out the brightness differences on the surface. Twelve bright regions have been identified on the planet, including a large north polar cap.

PHYSICAL DATA

SURFACE GRAVITY: = 0.8m/s^2 (low gravity causes the atmosphere to be more extended than the Earth's)

TEMPERATURE RANGE: = -233°C to -223°C

ATMOSPHERE: nitrogen, methane

SURFACE: rock and ice covered with a layer of frozen methane and nitrogen (similar to surface on Triton - a moon of Neptune)

Pluto has never been visited by a probe, so its exact characteristics are unknown, and given its distance from the sun, it is hard to make educated guesses. Pluto reflects quite a lot of sunlight, which has led scientists to predict that much of the planet's surface is covered in a layer of rock and ice. Scientists think the ice is mostly nitrogen, with some methane and traces of carbon monoxide. When Pluto makes its closest approach to the sun, inside the orbit of Neptune, the ice on the surface begins to melt and evaporate, giving rise to a nitrogen atmosphere. In 2002, when Pluto occulted a star, scientists were surprised to discover that the atmosphere appeared to be getting thicker, despite the fact that it is now moving outside Neptune's orbit and away from the sun. Scientists believe this unexpected activity might be caused by heat that Pluto may have stored during its closest approach to the sun.

THE KUIPER BELT

ABOVE: image of Sedna, a large Kuiper Belt object. This image suggests an upper limit on Sedna's size of 1600 kilometers in diameter. Sedna orbits the sun far beyond the ninth planet Pluto. Sedna was discovered by ground-based telescopes in 2004.

After years as an oddity in the solar system, Pluto found its true place as a member of the Kuiper Belt. The existence of a belt of objects extending beyond the orbit of Neptune was first suggested by the Dutch-born astronomer, Gerard Kuiper. But it is also attributed to the British astronomer Kenneth Edgeworth, and as such is sometimes referred to as the Edgeworth-Kuiper Belt. They both believed that beyond Neptune were a series of comet-like objects, which were home to short-period comets, such as Halley's Comet. Evidence began mounting for the existence of the Kuiper Belt in 1992 when the astronomer David Jewitt discovered a small planetoid beyond the orbit of Pluto. Since then, several hundred trans-Neptunian objects have been discovered, and there may be many thousands more. The objects in the Kuiper Belt form a band or ring around the ecliptic plane in which the planets orbit.

PHYSICAL DATA

DISTANCE FROM THE EARTH (MINIMUM):
= 4,290,000,000km

AVERAGE DISTANCE FROM SUN:
= 5,906,380,000km
(Earth = 149,597,000km)

AVERAGE SPEED IN ORBITING SUN:
= 5 km/sec (Earth = 30 km/sec)

THE PLUTO DEBATE

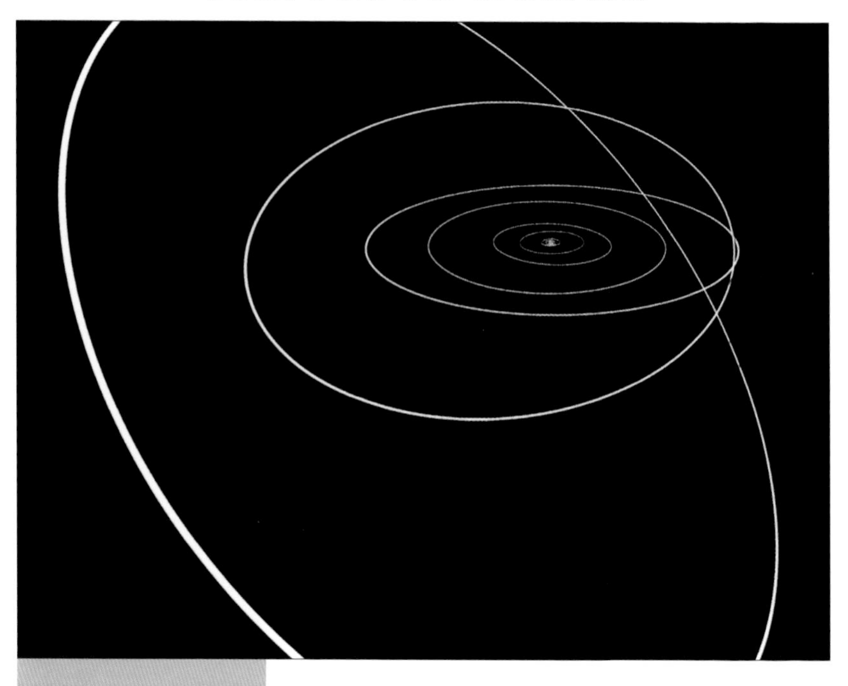

ABOVE: Computer graphic of the orbit (large oval) of what has been claimed as the tenth planet, 2003 UB313. The orbits of the other nine planets are also shown, the next planet being Pluto. Earth's orbit is shown in red close to the sun, which is at the center of the orbits. 2003 UB313 is a temporary name. It is thought to be larger than Pluto, and takes 560 years to orbit the sun. Its distance from the sun ranges from 38 to 95 times the Earth-sun distance.

Although the discovery of the Kuiper Belt put Pluto's role in the solar system into perspective, it remained the largest known Kuiper Belt object, meaning that it was able to cling on to planetary status. However, the announcement in 2005 that a trans-Neptunian object larger than Pluto had been discovered using photos from 2003, once again called Pluto's planetary status into question. 2003 UB313 is certainly larger than Pluto, although its exact size is not yet known. It is the most distant known solar system object and it is thought to have an orbital period of over 550 years. The International Astronomical Union needs to decide the fate of Pluto. If Pluto is to remain a planet, 2003 UB313 should also be given that credit, or both should be demoted to the status of minor planets or trans-Neptunian objects. The most likely outcome is that Pluto will be allowed to keep its planetary status for historical reasons, but not to designate that status to any further trans-Neptunian objects unless they are of considerable size.

NEW HORIZONS

More and more trans-Neptunian objects are discovered every year and this trend will continue in the future as telescopes are improved and as more scientists enter into this field of research. Pluto is the only planet never to have been visited by a probe. This will be rectified by the New Horizons spacecraft, which was launched in January 2006. The mission is hastily launched so that the probe might reach the planet before the atmosphere disappears. The probe will answer many questions presented by Pluto and Charon before moving on to look at other objects in the Kuiper Belt. If the probe detects something of interest, it can be maneuvered to intercept it. This mission will take several years to reach Pluto, after using Jupiter as a slingshot, but hopefully it will be worth the wait, giving us a new insight into this part of the solar system.

ABOVE: Illustration of Pluto and its companion Charon. The sun is visible at right. Pluto was first seen in 1930.

NEW HORIZONS:

The Atlas V rocket launched the New Horizons spacecraft on its 9-year journey to Pluto on 19 January 2006, at the Kennedy Space Center, Florida, USA. An additional solid propellant motor increased its speed until it became the fastest spacecraft ever launched, traveling at around 16 kilometers a second. It passed the Moon in nine hours, and will reach Jupiter in 13 months, using its gravity to swing past and accelerate towards Pluto. It is expected to reach Pluto in mid-2015 and spend 15 months studying the solar system's most distant planet.

ASTEROIDS, METEORS & COMETS

SIZE MATTERS

ABOVE: Image from the Mars exploration rover. This was the first meteorite to be found on another planet. It is about the size of a basket-ball and is composed mainly of iron and nickel.

PHYSICAL DATA

CERES

DIAMETER: = 1000km

MASS: = 8.7 x 10^{20}kg

ROTATION PERIOD: = 9 hours 5 minutes

PALLAS

DIAMETER: = 525km long

MASS: = 3.8 x 10^{20}kg

ROTATION PERIOD: = 7 hours 49 minutes

The key difference between asteroids and meteoroids is size. While asteroids are fragments of rock and metal which are greater than 50m in diameter, meteoroids are small fragments of rocks and minerals with diameters less than 50m. They are usually no larger than boulders in space and often even just the size of a grain of sand. If a meteoroid strikes the Earth it will burn up in the Earth's atmosphere, presenting a beautiful, harmless, spectacle – a meteor. However, if an asteroid were to strike the Earth, it could effect the extinction of mankind. Such extinction events are not just theoretical; it is now widely assumed that the asteroid which created a 300km basin in the Yucatan Peninsula in Mexico was responsible for the death of the dinosaurs.

THE ASTEROID BELT

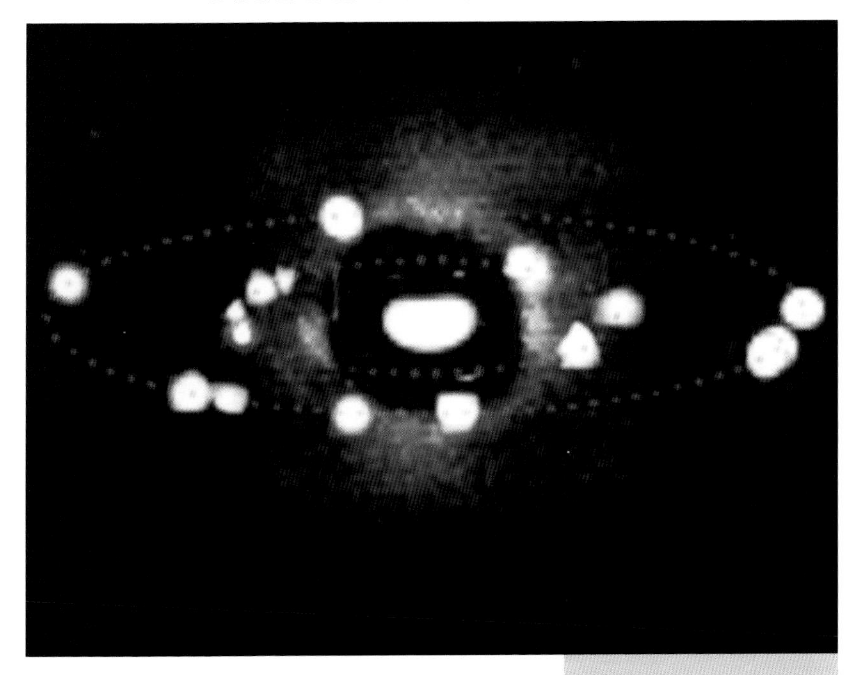

Many of the asteroids in the solar system are located in a band between Mars and Jupiter called the Asteroid Belt. Thousands of asteroids are known to exist in the Belt, and many thousands more are predicted. It is believed that the Belt is a failed planet – the chunks of metal and rock were unable to group together to form a fifth terrestrial planet because the process was interrupted by the strong gravitational pull of Jupiter. However, most of the asteroids are so small, that even if they had all united, the resulting planet would be even smaller than the moon. The first asteroid to be discovered, Ceres, was later found to be by far the largest asteroid in the Belt. After the discovery of the first asteroids at the start of the nineteenth century, William Herschel offered the term "asteroid," meaning "star-like." This highlighted the fact that, although asteroids moved like planets across the background stars, from Earth they looked more like stars than planets because they are too small to exhibit a disc.

ABOVE: Telescope image of Asteroid 87 Sylvia and its moons. Moons are shown in various positions around 87 Sylvia (center). Romulus, its outer moon, was discovered in February 2001. Remus, the inner moon, was discovered in August 2005.

PHYSICAL DATA

87 SYLVIA:

SIZE: = 260km long

ROMULUS (OUTER MOON):

SIZE: = 18km wide

ORBITAL DISTANCE FROM 87 SYLVIA: = 1360km

REMUS: (INNER MOON):

SIZE: = 7km wide

ORBITAL DISTANCE FROM 87 SYLVIA: = 710km

COLLISION COURSE

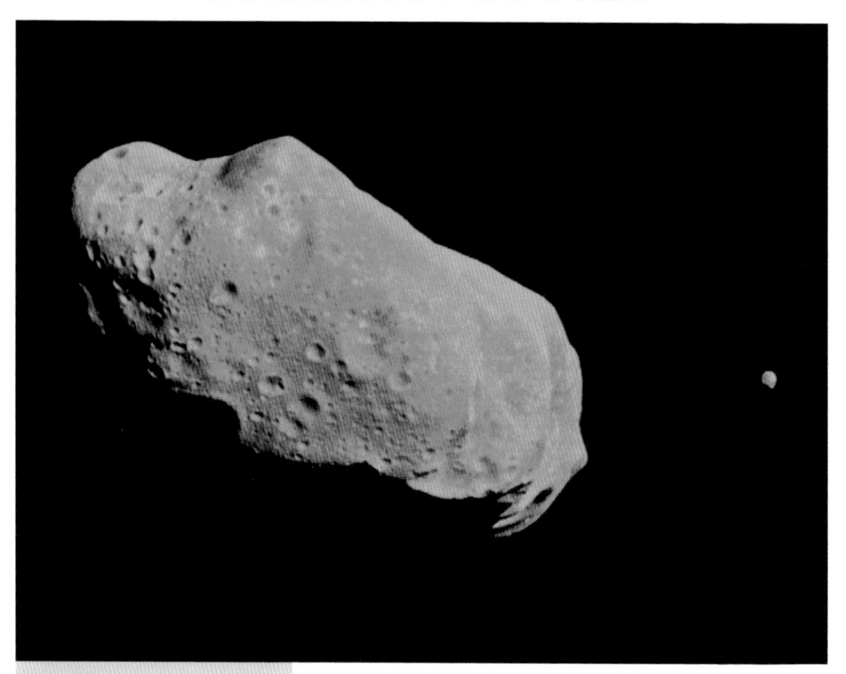

ABOVE: Colored image gathered by the Galileo spacecraft showing Asteroid 243 Ida and its tiny moon, provisionally known as 1993 (243)1. This picture was the first conclusive evidence that an asteroid can have a satellite. Ida is a member of the Koronis family of asteroids.

PHYSICAL DATA

IDA:

SIZE: = 56km long

MOON OF IDA (DACTYL):

SIZE: = 1.5km wide

ORBITAL DISTANCE FROM IDA: = 100km

The orbits of asteroids are not just confined to the Asteroid Belt, rather they can be found throughout the solar system. Trojan asteroids can be found in the orbital paths of Jupiter, Neptune and Mars, and a number of trans-Neptunian objects in the Kuiper Belt can also be classified as asteroids. However, some follow highly elliptical orbits of the sun – these are often comets which have lost their volatile materials. These asteroids intersect Earth's orbital path during their solar orbit, and are classified as "near-Earth asteroids." If an asteroid and the Earth happened to be at the same point of their orbital paths at the same time, the result would be a collision. Scientists are constantly monitoring for such a collision event. The asteroid with the greatest chance of impacting the Earth is 1950DA, which might hit the planet in March 2880. However, such estimates are constantly revised.

PROBING ASTEROIDS

During the Cold War, sending probes to asteroids was not at the forefront of either superpower's space programs, largely because asteroids lacked the prestige element offered by missions to other planets. An asteroid was not visited by a probe until the Galileo probe flew past Gaspra in 1991, and that was a flying visit while en route to Jupiter. Five years later, NASA launched its NEAR (Near-Earth Asteroid Rendezvous) probe which was the first destined exclusively for an asteroid, in this case Eros. The probe spent a year mapping Eros, gaining basic understanding of its size, shape and surface temperatures. There was more glory for NASA when NEAR successfully landed on the surface of the asteroid, something that had not been intended prior to its arrival. NASA's Dawn Mission will be launched towards the Asteroid Belt in summer 2006. The probe is designed to reach Vesta in 2011, where scientists hope to find indications of volcanic activity, before flying on to Ceres in 2015.

ABOVE: NEAR (Near-Earth Asteroid Rendezvous) spacecraft image of Asteroid 253 Mathilde. It is the commonest type of asteroid. Its surface is marked by many large impact craters; the shadowed crater is about 10km deep.

PHYSICAL DATA

MATHILDE

SIZE: = 59 x 47 km

DISTANCE FROM SUN: = 290 - 500 million km (2 - 3.5 times the radius of the earth's orbit)

ORBITAL PERIOD: = 4.3 years

ROTATIONAL PERIOD: = 17 days

SHOOTING STARS

ABOVE: A meteor enters the Earth's atmosphere with a speed of up to 100 kilometers per second. Air resistance heats the particle, making it visible as a streak of light. Another meteor is seen at upper left.

TIMELINE

1801 - Italian astronomer, Giuseppe Piazzi discovered the first and largest asteroid, Ceres

1802 - German astronomer, Wilhelm Olbers discovered the second asteroid, Pallas

1804 - K. Harding discovered the third asteroid, Juno

1991 - First visit to an asteroid (Gaspra) by the probe Galileo while en route to Jupiter

1997 - Hubble Space Telescope mapped Vesta (third largest asteroid) and found an enormous crater formed a billion years ago

When meteoroids collide with the Earth they burn up in the upper atmosphere, creating a striking spectacle called a meteor. Energy in the form of heat is generated as the meteoroid travels at supersonic speeds, compressing the air beneath it. This energy strips the meteoroid of its electrons in a process is called ionization, which leaves a trail in its wake. This ionization trail glows momentarily as a streak through the sky, earning meteors the common name, "shooting stars." Occasionally, a meteor is particularly bright; this means the meteoroid is larger, probably coming from an asteroid, or even the moon or Mars. These meteors are named "fireballs" rather than shooting stars. They are usually accompanied by the sound of a sonic boom and, if they reach the surface, usually result in an impact crater. Small meteoroids collide with the Earth with great frequency.

METEOR SHOWERS

ABOVE: Optical time-exposure image of Leonid meteors (streaks) against a starfield containing the Milky Way (band across center). The Leonid shower occurs each year for about two days around 17 November, when the Earth crosses the debris produced by the comet Tempel-Tuttle.

As comets orbit the sun they leave meteoritic debris in their wake. If the comet is a short-period comet, then this debris will, over time, spread evenly throughout the comet's orbital path. When the Earth intersects the orbital path of a comet the debris rains down on Earth as a stream of meteors called a "meteor shower." The most famous meteor showers are the Leonids, from the comet Tempel-Tuttle, which occurs in mid-November, and the Perseids, associated with the orbit of comet Swift-Tuttle, which peaks on 12 August every year, but is visible during late July and early August. More than fifty meteor showers occur annually; although many are very weak, there are several stronger showers to look out for, such as the Geminids and the Ursids in December, the Orionids in mid-October and the Quadrantids at the start of the New Year.

METEOR SHOWERS

NAME	COMET	MONTH
Leonids	Tempel -Tuttle	November
Perseids	Swift -Tuttle	August
Geminids		December
Ursids		December

METEORITES

While most meteoroids that collide with the Earth are small and disintegrate in the upper atmosphere, some are able to reach the Earth. When part of a meteoroid reaches the surface it becomes known as a meteorite. When meteorites strike the Earth they often leave an impact crater. Before the Space Age, meteorites were the only way that material from outer space could be handled and analyzed. While most meteorites come from asteroids or comets, rarely they prove to have come from the moon or Mars. Surface rock and metals become dislodged when both are themselves impacted by meteorites and these fragments sometimes manage to escape into space and after many years they might fall to Earth and land as meteorites. While all meteorites are precious, lunar and Martian meteorites are so rare that they are an exceptional find. Fewer than 100 lunar meteorites have been discovered, and only 34 meteorites of Martian origin have been confirmed. The first known Martian meteorite to fall to Earth fell in Chassigny, France in 1815, although its Martian origin was not known at the time. The meteorite Yamato 791197, discovered in 1979, is the first known lunar meteorite.

LEFT: Aerial view of Meteor Crater, Arizona. Sometimes called the Barringer Crater, after the mining engineer who first suggested that it was formed by a meteor impact, it is believed to be about 50,000 years old. The crater is in northeastern Arizona, near Winslow, and is about 200 meters deep and 800 meters across.

GIANT SNOWBALLS

HALLEY'S COMET

ABOVE: Halley's comet, photographed from the Royal Observatory, Cape of Good Hope, 1910. Named after Edmund Halley (1656-1742), who first recognised its periodicity, the comet has been recorded every 76 years since 240 B.C.. It consists of a potato-shaped nucleus of ice and dust. Its tail is formed as the comet is heated when it approaches the Sun. Dust and gas are released and swept into a tail both by radiation pressure and by the solar wind.

In spite of their sheer beauty, many civilizations throughout history have considered the arrival of a comet as a prophecy of doom. Understandable perhaps, given that comets did not seem to conform to the movements of other objects across the night sky, appearing at what seemed like random occasions. Nowadays, comets are much better understood. At their heart is an asteroid-like object called the nucleus, which contains rock, frozen gases and water, meaning that comets are like giant snowballs. The nucleus is obscured by the coma, a region of gas and dust lifted from the nucleus as it begins to near the sun. As the comet gets even closer to the sun, material in the nucleus begins to evaporate. The evaporated material is streamlined by solar winds, creating the comet's tails; comets usually have two – a dust tail, and an ion tail. The dust tail is composed of the evaporated ice and dust fragments extending several million kilometers behind it. The ion tail is even longer, although it is less easily visible in the night sky because of its darker, bluish color. It results when gases from the coma react with the sun's rays and undergo a process of ionization.

ORBITAL PERIODS

ABOVE: The nucleus of Comet Halley seen by the Giotto spacecraft on the night between the 13 and 14 March 1986. It has a potato shape with dimensions 8.2x8.4x16km and a mass of about 100 billion tons. On the left edge of the nucleus active sites are clearly visible. Here the water ice, heated by sunlight, sublimates to form an atmosphere, also known as the coma. The escaping gases sweep up particles of dust which are carried from the nucleus into the coma. These particles form the comet's tail.

The fact that many comets periodically repeat their orbits of the sun was not known until 1705, when Edmond Halley realized that the comet later to be named in his honor was not several different comets appearing randomly, but rather was the same body on a repeat orbit of the sun.

Three classifications of comet can be identified by their type of orbit. The Jupiter family of comets follow a path between the sun and Jupiter, which means that they orbit the sun frequently, taking around five to ten years. The second group are short-period comets, including Halley's Comet, they have the Kuiper Belt as their aphelion (their furthest point from the sun). The third group are long-period comets which take many years to orbit the sun as they return to the Oort Cloud at the very edge of the solar system. Most long-period comets are never noticed by humans because they orbit the sun at such a great distance that even our best telescopes cannot detect them. The Oort Cloud is so far away from the sun that it is sometimes influenced by the gravity of other stars, which can knock a comet out of its orbit within the Oort Cloud and send it hurtling into the known solar system, where it can be detected and cataloged.

DEEP IMPACT

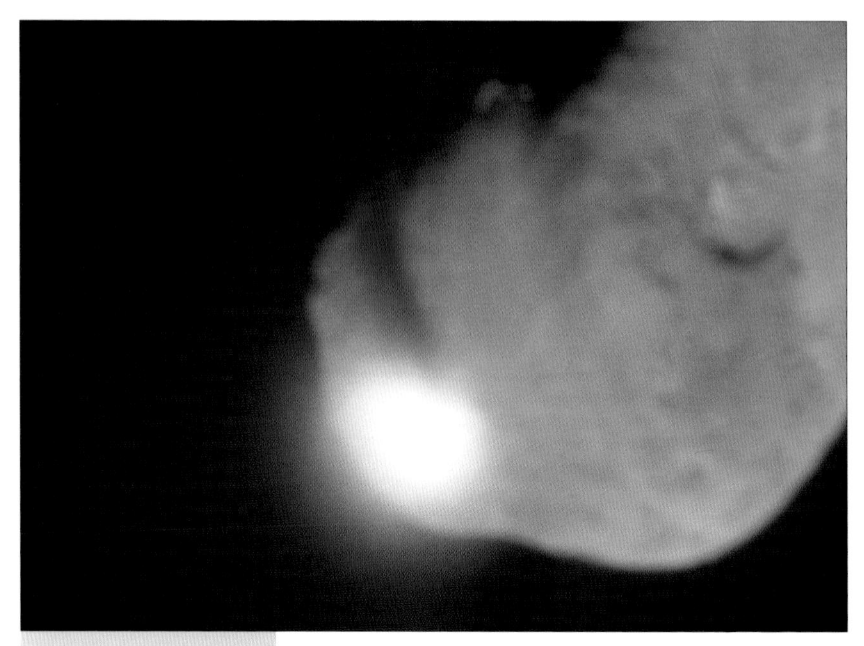

ABOVE: Deep Impact comet strike. Explosion (white) on the surface of the nucleus (gray) of the comet 9P/Tempel 1, after being struck by the Deep Impact impactor. The impact caused the ejection of a large amount of interior material.

In 2005, NASA used its Deep Impact Probe to investigate the internal makeup of a comet. It is thought that some of the ice water found on comets might be responsible for water on Earth, following impacts early in the planet's history. Moreover, it is believed that comets contain material that has not been changed since the creation of the solar system; a look inside a comet might give us a glimpse back to those days at the dawn of the solar system – the chosen comet, Tempel 1, was turned into a 4.6-billion-year-old time capsule. The probe's impact with the comet caused a long plume of material similar in consistency to talcum powder to burst out from the comet. It is expected that analysis of the mission's findings will give us a much greater insight into the internal structure of a comet's nucleus. The results of Deep Impact will be corroborated by a European mission named Rosetta, which blasted off in 2004 on a journey to the comet Churyumov-Gerasimenko. It is due to set down a lander on the comet in 2014.

HALE-BOPP

In June 1995, the comet Hale-Bopp was discovered independently by Alan Hale and Thomas Bopp. It became visible to the naked eye during the latter half of 1996, disappearing as it neared the sun at the end of that year. When it reappeared in early 1997, it was a spectacular sight, lasting the entire year. It was easily visible in the sky, even in areas obscured by light pollution. Its longevity made Hale-Bopp spark global interest among astronomers and the general public alike. Astronomers watching it carefully even detected a third tail, comprising sodium, on Hale-Bopp. The last naked-eye observations of the comet were made in December 1997, after a total of more than eighteen months' visibility, which doubled the record for the longest period that a comet had been visible to the unaided eye.

ABOVE: The comet Hale-Bopp showing its gas and dust tails. Hale-Bopp was one of the brightest comets of the 20th century. Its gas or "ion" tail (blue) consists of ionized glowing gas blown away from the comet head by the solar wind. The dust tail (white) consists of grains of dust pushed away from the comet head by the radiation of sunlight. Comets have a nucleus of ice and dust. As they approach the sun their surface evaporates, releasing a tail millions of kilometers long. Comet Hale-Bopp was discovered on 23 June 1995. Photographed on 6 April 1997 near Wesel, Germany.

COMET HALE-BOPP PHYSICAL DATA

DIAMETER: estimated 40km

ROTATION PERIOD: 11 hours 24 minutes

LENGTH OF TAIL: 85 - 90 million kilometers

CLOSEST DISTANCE FROM EARTH: 196,000,000km

STARS

BIRTH OF A STAR

ABOVE: Satellite image of the red dwarf Proxima Centauri (center). Explosive outbursts (flares, red) occur almost continually in the star's upper atmosphere. These flares are caused by the star's low mass, which is a tenth that of the sun's. Nuclear fusion reactions that convert hydrogen to helium proceed very slowly in Proxima Centauri, creating a turbulent, convective motion throughout the star's interior. This motion stores up magnetic energy, which is then released explosively. Image taken on 7 May 2000 by the Advanced CCD Imaging Spectrometer (ACIS) aboard NASA's Chandra X-ray Observatory.

Interstellar hydrogen gas is not found uniformly throughout the universe; instead, it is found in patches because gravitational attraction pulls it together. These patches are molecular clouds, a cradle in which new stars are formed. After millions of years some, as yet unknown, force disrupts the large cloud causing it to break up into smaller clumps of gas, from which a proto-star is born. The proto-star is a star's formative phase, during which it strives to gain equilibrium between its internal forces and gravity. Proto-stars are usually difficult to detect optically because they are shrouded by thick layers of gas and dust. Proto-stars can last between 100,000 years and 10 million years. During this period, the proto-star spins very rapidly, generating intense heat and pressure causing the cloud to collapse further, meaning more hydrogen is accumulated in the core of the proto-star. Eventually the temperatures and pressures in the star are sufficient that hydrogen fusion can begin, and the star is born.

PROTO-STAR

Before fusion begins, the proto-star contracts because of the gravitational pressure exerted upon it. Initially a proto-star can be vast; some are billions of kilometers in diameter. Once fusion begins, the star starts to emit electromagnetic radiation, which counterbalances the force of gravity and causes the star to begin expanding. When the star becomes too massive, the radiation is unable to overcome the force of gravity, which regains the ascendancy and the star contracts until its surface area is of such a size that the force of the radiation is once again greater than the force of gravity and the process repeats itself. This to-ing and fro-ing continues for many years until the star finally reaches equilibrium, when gravity, the external force, is balanced by internal forces. This unstable period is called the star's T Tauri phase, because the star T Tauri was undergoing this process when it was discovered. It can be detected because the star's output of energy fluctuates significantly over time.

ABOVE: Star trails over a lake. These streaks of light are formed on long-exposure photographs due to the apparent motion of the stars caused by the Earth's rotation.

PREVIOUS SPREAD: Splendid starfield centered on the constellation of the Southern Cross (Crux Australis). It shows an area of our galaxy, the Milky Way, extremely rich in star clusters, dark and bright (pink) nebulae. Four bright stars at upper center identify the familiar cross-shaped asterism in the Southern Cross.

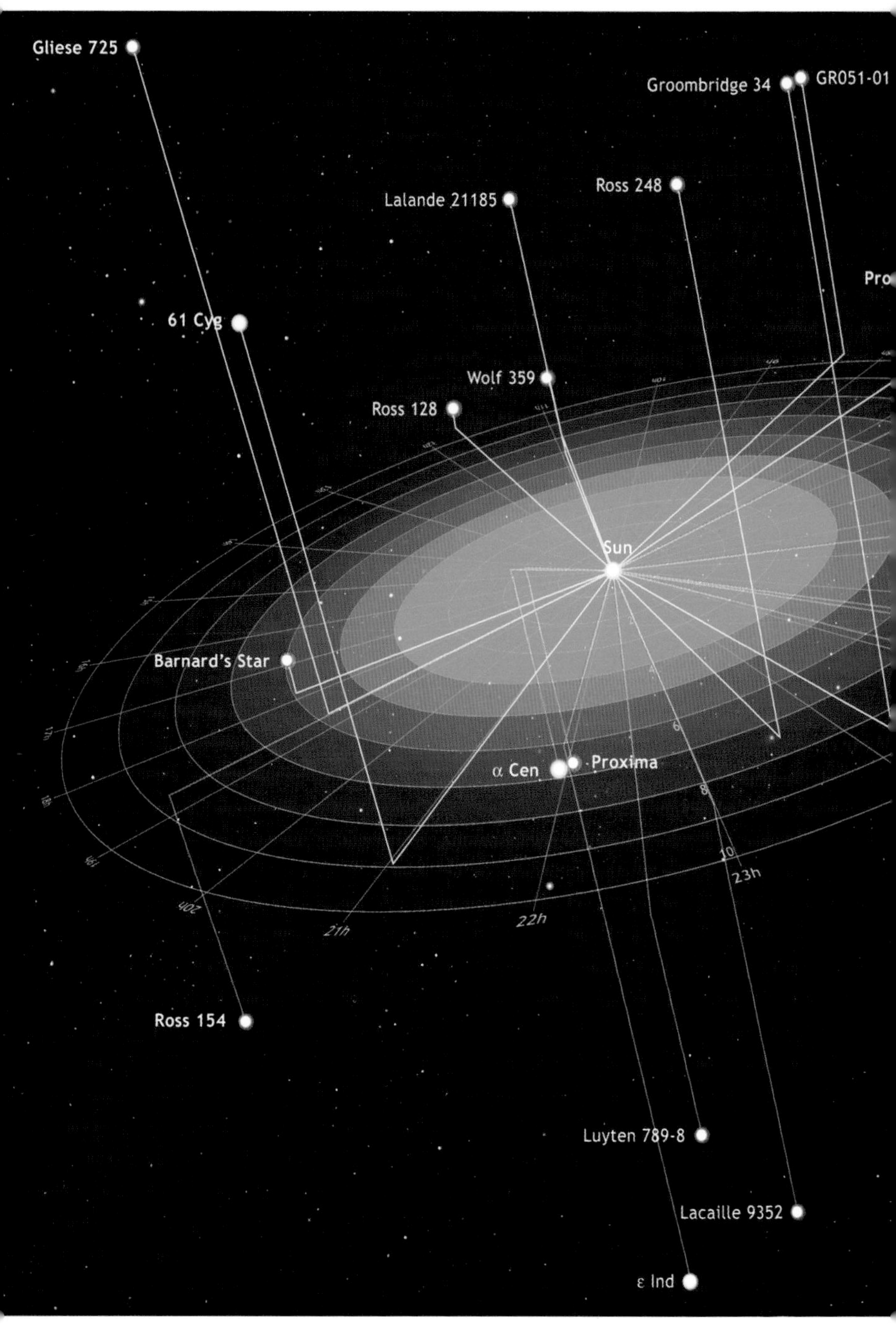
Gliese 725
Groombridge 34
GR051-01
Ross 248
Lalande 21185
Pro
61 Cyg
Wolf 359
Ross 128
Sun
Barnard's Star
α Cen
Proxima
21h
22h
23h
Ross 154
Luyten 789-8
Lacaille 9352
ε Ind

HYDROGEN FUSION

Once temperatures inside the core are hot enough, hydrogen fusion begins. In this process, hydrogen nuclei are fused together to make helium atoms. When one such atom is created, a small amount of energy is released. In a stellar core, this occurs millions of times each second, meaning that incredible amounts of energy are generated. When two scientists, Henry Russell and Ejnar Hertzsprung, working independently, put this information on a diagram, plotting temperature against luminosity, it emerged that the stars which produced their energy by hydrogen fusion were all banded together at the center of the graph. This band was called the Main Sequence, and all stars in their hydrogen fusion phase belong to it. Most of the stars we see in the celestial sphere are on the Main Sequence. It is the longest and most stable period of a star's life.

The mass of a star determines how long it will remain on the Main Sequence; the most massive stars tend to live fast and die young, rapidly fusing their hydrogen supply, while less massive stars tend to take a long time to complete their process of hydrogen fusion.

The mass of a star has implications for its temperature as well as its life expectancy. While the least massive stars have temperatures less than 30,000 degrees C, that of the most massive stars can be in excess of 300,000 degrees C. Our sun has a relatively low mass, with core temperatures around 60,000 degrees C. The temperature of a star influences its color. The hottest stars glow blue and white, while the coolest stars appear red. Stars with moderate temperatures, including our sun, are yellow and orange.

OPPOSITE: The 30 stars nearest the sun (center) are found in the 20 star systems shown here. The concentric circles are one light year apart, out to ten light years from the sun, and lines show the positions of the stars above (yellow) or below (red) the celestial equator. The closest stars are part of the Alpha Centauri star system (below sun). The brightest star in the night sky, Sirius, is at center right. The colors of the stars indicate the type of star (sun-like stars are yellow).

FAILED STARS

ABOVE: Galactic Center and Galactic Dark Horse, the Milky Way from Serpens to Scorpius. The Milky Way has a mass of roughly 750 billion to one trillion solar masses and a diameter of about 100,000 light-years.

Sometimes the proto-star never becomes hot enough to undergo hydrogen fusion. Such failed stars are called brown dwarfs because of their dull red-brown color. The temperatures and pressures never reached high enough levels in these stars to initiate the process of nuclear fusion in the core. Brown dwarfs can only be a maximum of 80 times more massive than Jupiter; any greater, and the mass would be such that fusion can occur. While stars do not emit most forms of radiation until the fusion process has begun, infrared radiation is emitted before this stage, while the star is first contracting under the weight of gravity. It is infra-red radiation that has allowed scientists to discover the existence of brown dwarfs, because they are almost impossible to see optically. The first discovery of a brown dwarf, Gliese 229B, was relatively recent, in 1995. Since then, many more have been found. However, the existence of a brown dwarf is usually difficult to confirm because stars are often found in orbiting pairs, called binaries, and there is some controversy over whether a contender is indeed a brown dwarf or if it is just a very large planet. The key difference is in how each object was formed; stars condense out of the interplanetary nebulae, while planets are formed by the accretion of smaller particles which form around those new-born stars.

DYING STARS

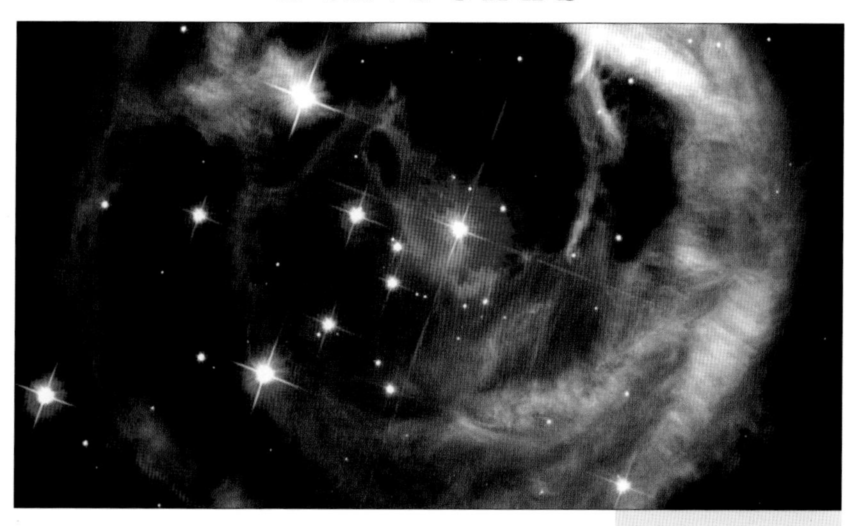

When a star's core stops fusing hydrogen to make helium, it exits the Main Sequence and enters its death throes. The equilibrium the star fought so arduously to attain at the beginning of its life is upset, as the radiation output ceases and the star begins contracting under gravity. The energy generated by this collapse heats up the stellar atmosphere where hydrogen is still present. The stellar atmosphere becomes so hot and pressurized that fusion of this hydrogen store takes place. This process causes the star to swell in size to become a giant. The increase in surface area of a star causes it to cool, and therefore redden. As such, most giant stars are red giants. The third brightest star in the sky, Arcturus, is a red giant. It is expected that our sun will become a red giant in about 5 billion years' time. If the temperature in the atmosphere of the red giant is hot enough to begin hydrogen fusion, the temperature in the core must be even hotter and under even greater pressure. Core temperatures in giant stars are in excess of 100 million degrees Kelvin. At such temperatures, the nuclear fusion of helium can occur. Helium fusion results in the creation of carbon and oxygen.

ABOVE: Light echoes from an exploding star. Hubble Space Telescope image of an illuminated dust shell around the star v838 Monocerotis. This star underwent a massive brightening in January 2002, temporarily becoming the brightest star in the galaxy. The light from this outburst reflects from a series of dust shells around the star, which are thought to have been ejected during previous activity. The phenomenon allows study of the fine structure of the dust shells, which could help explain why the star behaves as it does.

PLANETARY NEBULA

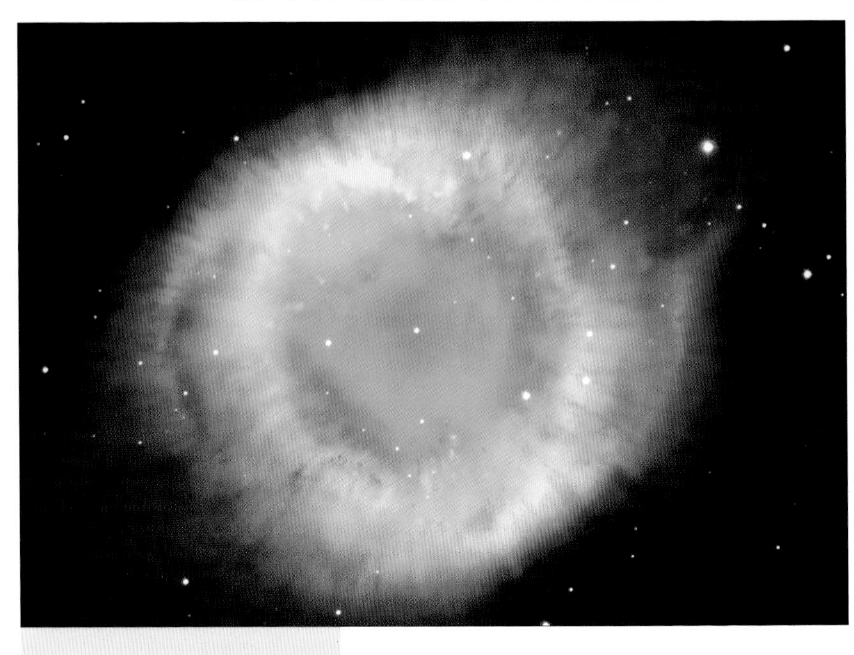

ABOVE: Enhanced optical image of the Helix planetary nebula (NGC 7293). This is the nearest planetary nebula to Earth, lying about 450 light-years away in the constellation Aquarius. Planetary nebulae are poorly named, as they have no association with either planets or nebulae. Instead, they consist of shells of gas ejected from the surface of a dying star. This is ionized by radiation from the star, causing it to glow. Their erroneous name arose from their appearance through early telescopes.

The process of helium fusion lasts for several billion years until the helium has run out. Core temperatures in giant stars never get hot enough to fuse the carbon and the oxygen. These stars stop producing energy and instead expel their remaining matter and radiation into space. This process manifests itself across the electromagnetic spectrum. Gamma rays, X-rays, ultraviolet and infrared radiation stream out and can be detected by scientists here on Earth. However, these invisible forms of radiation are not the only way to detect that this process is occurring, because it also manifests itself on the visible portion of the spectrum as a visually stunning planetary nebula. The name "planetary nebula" can be misleading; William Herschel coined the phrase because he thought these spectacles were nebulae resembling the discs of planets. It was only later discovered that these phenomena were dying stars.

WHITE DWARFS

ABOVE: Death of the sun. Artwork of the sun forming a planetary nebula 5 billion years from now. Ejected gas (purple) is expanding towards the lifeless Earth (lower right) and moon (lower left) from the dying sun (upper center). This stage in the life of the sun follows its expansion during its red giant stage. When its nuclear fuel has been used up, its outer layers are ejected in this way. The remnant cools to form a white dwarf star.

Once the star has ejected its matter, only a core comprising carbon and oxygen remains. The core continues to shrink under the weight of gravity. The contraction will eventually halt because the atoms are as crushed as they can be; electrons are pushed right up against the nucleus and begin repulsing one another. These relic stars are named white dwarfs. They are exceptionally dense, because so much matter is squeezed into such a small surface area. Any remaining energy and heat is exhausted by the white dwarf so that it reddens in color and will eventually "disappear" – inasmuch as it will no longer be detectable to us; a body of mass will still exist. As the process in which white dwarfs disappear takes billions of years and the universe is a maximum of fifteen billion years old, even the oldest white dwarfs are still detectable

SUPERNOVA

A supernova is a stellar explosion which is much brighter and more intense than a nova. They are so bright that they are easily visible from Earth. Supernovas are formed in two different ways, and therefore they are banded into Type I and Type II categories. Type I supernovas occur in similar circumstances to a nova. Novas occur when a white dwarf is orbiting another star in a pair, called a binary. The gravitational pull of the white dwarf can drag hydrogen from the partner star onto its shell. This new hydrogen supply is fused very rapidly causing a short-lived, extremely bright burst of light. When it runs out of hydrogen the star is once again extinguished and the process repeats itself. However, if a white dwarf accretes an unsustainable amount of hydrogen gas from its partner star, it begins to collapse and the white dwarf is destroyed in a mighty explosion, a supernova.

When stars exit the Main Sequence, having burned all their hydrogen, we have seen that they turn into red giants and undergo the fusion of helium into carbon and oxygen. However, stars which were already giants while they were on the Main Sequence are an exception to this trend. They swell up so greatly that they become supergiants. This means that not only are they able to fuse helium but subsequently also the carbon and oxygen. Oxygen fuses to create neon, which in turn fuses to create magnesium; the process continues and forges silicon from magnesium and then iron from silicon. The fusion process does not create any more energy by the time an iron core has been reached, so the iron is not fused. When the core of a supergiant star turns to iron, it has reached the end of the road. The star collapses under the colossal gravitational forces at play on the heavy iron core. This swift collapse results in an immense explosion: a Type II supernova, as the star ejects its material in a spectacular shockwave. The energy produced during a supernova is so great that fusion of iron can finally begin; thus, it is only because of supernovas that we have all the elements heavier than iron.

The supergiant Betelgeuse in the Orion constellation will inevitably become a Type II supernova. It is the ninth brightest star in the night sky, so the supernova will be a remarkable sight. Some scientists believe that the star may already be in its carbon fusion stage, meaning that the star could become a supernova in less than one thousand years.

OPPOSITE: Optical image of the Crab nebula, also known as Messier 1. The nebula is the remnant of a supernova explosion observed in 1054 AD, and consists mainly of hydrogen gas. The nebula is 6000 light-years from Earth in the constellation Taurus.

NOTABLE SUPERNOVAS

1054: Chinese Astronomers note a supernova in the constellation Cassiopeia.

1572: "Tycho's Star" - the astronomer, Tycho Brahe observed another supernova in the constellation Cassiopeia. It was as bright as Venus in the sky and even visible during daylight.

1604: "Kepler's Star" - Brahe's colleague Johannes Kepler first spots a supernova in the constellation Ophiuchus. It was the last recorded supernova in the Milky Way.

1987: A supernova in another galaxy, the Large Magellanic Cloud, could be seen with the naked eye in the southern hemisphere.

NEUTRON STARS

ABOVE: Artwork illustrating the concept of warped space. The fabric of space is represented as a grid in which objects of increasing mass produce increasingly large distortions. Our sun, at bottom left, makes almost no impression. A small, but much denser and more massive neutron star (lower center), creates a slight distortion. The enormous gravitational pull of a black hole (bottom right) creates a yawning chasm, warping the fabric of space for light years around. The idea that space is distorted by gravity is a consequence of Einstein's Theory of Relativity. The grid of space is like a thin rubber sheet on which objects of varying weight produce smaller or larger dents.

After a supernova all that remains is the star's core. What becomes of the core depends upon its mass. Cores with a mass of between 1.4 and 3 times that of our sun become neutron stars. The figure of 1.4 solar masses was calculated by the Indian astronomer Subrahmanyan Chandrasekhar in 1930, and is known as the Chandrasekhar Limit. In a white dwarf, the core cannot collapse any further because of the electron repulsion. However, if a star is more than 1.4 times as massive as our sun, gravity would be sufficient to overcome this hurdle. The negatively-charged electrons are pushed into the nucleus where they collide with the positively-charged protons to leave a neutrally-charged core of neutrons. There is insufficient gravity for stars to collapse further, the outward force of the neutrons balances the inward force of gravity. This star, with its neutron core, is unimaginatively named a neutron star. These stars are exceptionally small, averaging just 20km in diameter. A mass at least 1.4 times as great as our sun is condensed into such a small area, making neutron stars the densest objects in the universe.

BLACK HOLES

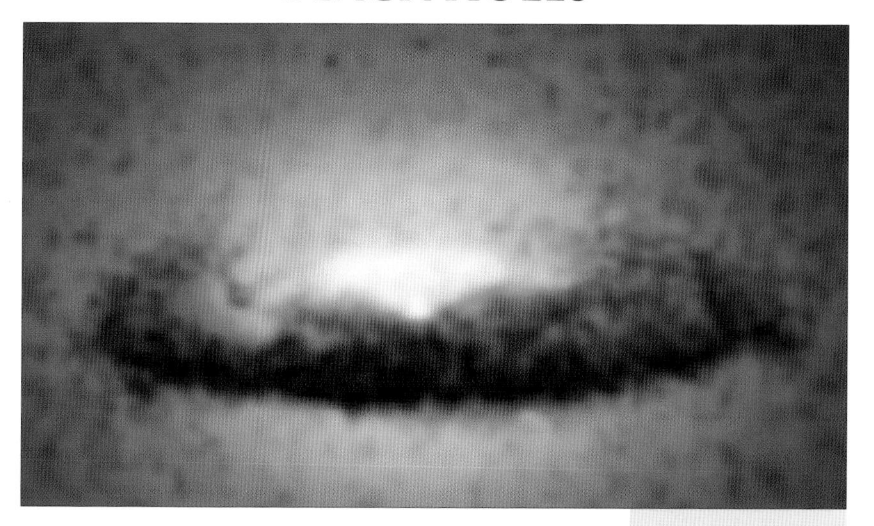

ABOVE: Hubble Space Telescope optical image of a massive black hole in the center of the elliptical galaxy NGC 7052. Surrounding the galaxy is a 3700 light-year (LY) diameter dust disc (brown). The bright spot (white, center) is light from stars crowded around the black hole due to its powerful gravitational pull. This black hole is about 300 million times the mass of our sun. The disc rotates at 341,000 miles per hour; the dust may come from an ancient galaxy collision. NGC 7052 in the constellation Vulpecula is 191 million light-years from Earth.

If the star undergoing a gravitational collapse is three or more times as massive as our own sun, gravity is so strong that the collapse is unstoppable, even the neutrons are crushed. The result is a black hole. The mass of a massive star collapses into a point without any volume, called a singularity. With such a large mass in an infinitely small area, the gravitational force of the black hole is so strong that the escape velocity required for an object to overcome it is greater than the speed of light (300,000,000 meters per second). The speed of light is the maximum attainable speed in the solar system, nothing can exceed it, meaning nothing can escape a black hole. As no light can escape, this phenomenon can not be detected optically (or indeed by any other emissions on the electromagnetic spectrum), leading the American physicist, John Wheeler, to coin the term "black hole" in 1967.

CONSTELLATIONS

For the ancients the importance of observation of the heavens cannot be underestimated. It was the means by which the cycles of the year and longer-term phases in time could be measured, calculated and foretold. Astronomers in all early civilizations saw patterns in the stars by which time and place could be fixed. Inevitably perhaps, the eternal and apparently unchanging nature of the skies meant that they became closely associated with, or indeed identified as, these civilizations' gods. Thus stars and star patterns were incorporated into mythologies concerning the exploits of deities and their interactions with humanity, so that many of the constellations we can view today are named after beings and objects in those ancient tales.

The stars' relationship with gods and their unchanging nature led also to a belief that somehow they controlled the fate of individuals and societies. In the past, many who we would today class as astronomers, and thus scientists, would also practice astrology – trying to divine the future from the positions of stars and constellations. The most well-known astrological symbol is the zodiac, a calendar of constellations which appear in the skies of the northern hemisphere as the year progresses; it was believed (and still is by some people) that the dominant constellation at the time of birth would determine the character and fate of an individual.

In the west, learning and knowledge developed from Greek and Roman civilizations and this is where many of the internationally acknowledged names for stars and constellations have their origins. What follows here is just a taste of some of those constellations, describing where, and when, they can be located in the night sky. Most are visible from the northern hemisphere, but some, like Crux, or the Southern Cross, can only be seen from the southern hemisphere. It is important to remember when looking at individual stars in a given constellation that, while from the Earth they are in the same region of the sky, they may be millions of light-years distant from each other.

AQUARIUS

THE WATER-BEARER

ABOVE: Aquarius is the eleventh of the twelve constellations of the zodiac. It is an extensive southern hemisphere constellation. The whole area of the sky in which it lies was associated in ancient times with water. The Babylonians called the area the Sea, and populated it with ocean creatures such as Cetus, Pisces, Capricornus, Delphinus - all controlled by Aquarius. The Egyptian hieroglyph for water is the same as the symbol used for Aquarius.

ARIES

THE RAM

ABOVE: Aries is the first of the twelve signs of the zodiac, because in ancient times it was where the sun's path crossed the celestial equator. Since then this point has passed into Pisces due to the Earth's wobble (precession). Aries has been associated with the ram for at least 2000 years. The Greeks associated Aries with the story of the Argonauts and the Golden Fleece. The Chinese called it the Dog, part of a larger figure that included Taurus and Gemini.

CRUX - THE SOUTHERN CROSS

(Previous Pages)

The dark patch just below center is called the Coal Sack. It is a nearby dark nebula which obscures the light of the bright Milky Way behind it. The bright star at the left of the cross, just above the Coal Sack, is Mimosa (Beta Crucis). To the right of the Coal Sack, at the foot of the cross, is Acrux (Alpha Crucis). The bright star at far left is Hadar (Beta Centauri) in the neighboring constellation Centaurus, the centaur. Crux is best seen in the the fall from the southern hemisphere.

GEMINI

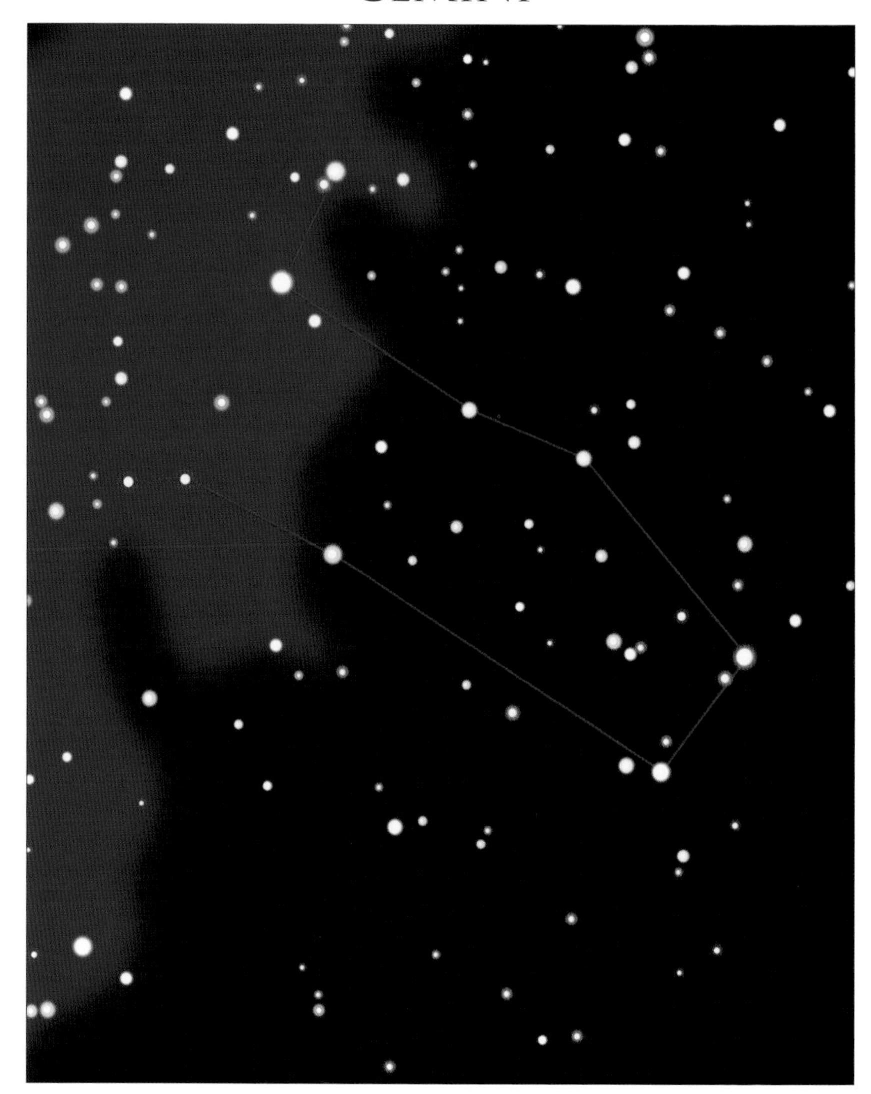

THE TWINS

ABOVE: The twins are the two brightest stars in the lower right of the constellation, with Pollux above and to the right of Castor. In Roman mythology they were the sons of Leda and the Swan who was Zeus disguised. Born of the same egg, they were placed in the heavens because of their brotherly love, although they represent the opposing principles of war (Castor) and peace (Pollux). They were also the guardians of Rome and its sailors.

LEO

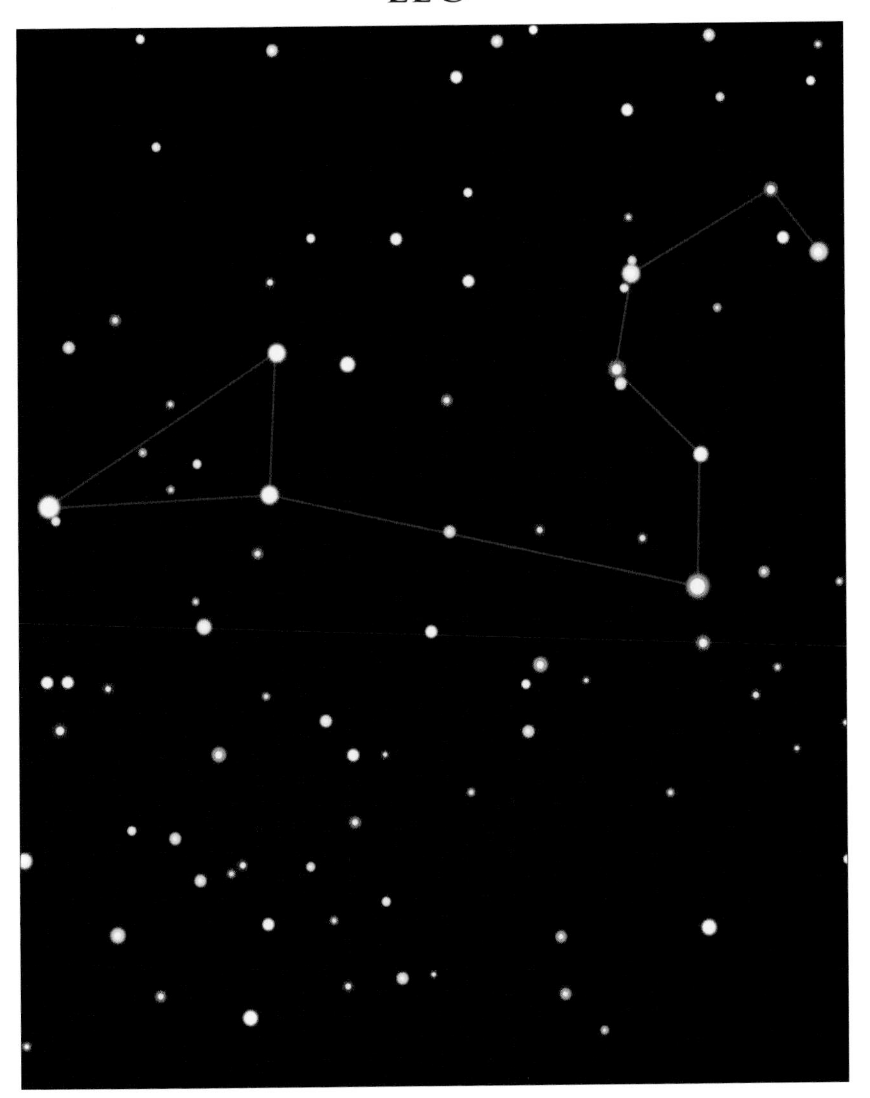

THE LION

ABOVE: Leo is the fifth of the twelve signs of the zodiac. The stars of Leo form two groups, a triangle forming the lion's haunches and tail, whilst a sickle-shape forms its head and mane. Leo's brightest star, Regulus, is at the base of the sickle and seems to be more important in mythology than the lion itself. The Egyptians accorded it power because the annual Nile flood came at the time when the sun entered Leo.

ORION

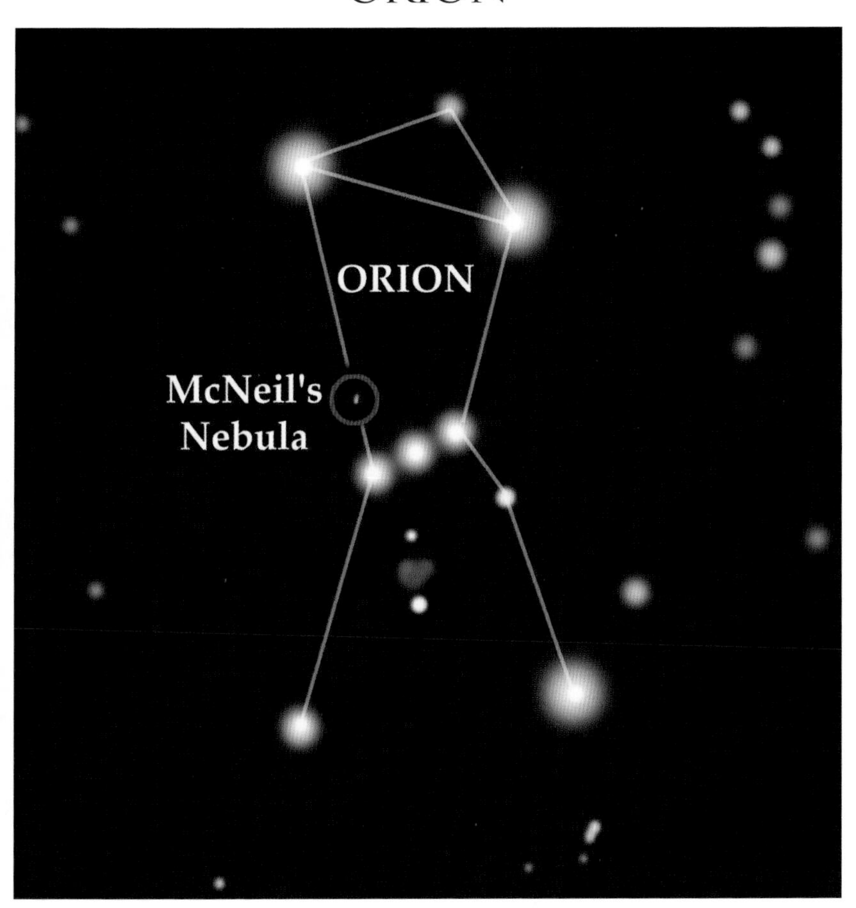

OPPOSITE: The most prominent feature of the Orion constellation is Orion's Belt, a row of three bright stars (center left). From left to right, these are Alnitak, Alnilam and Mintaka (Zeta, Epsilon and Delta Orionis respectively). At upper left is the red supergiant star Betelgeuse (Alpha Orionis). The blue supergiant star Rigel (Beta Orionis) is at lower right. Directly below the belt, the Orion nebula is seen as a pink smudge. Orion's Belt straddles the celestial equator, so the constellation is seen equally well from both hemispheres.

ABOVE: Diagram of stars and other objects in the Orion constellation. This is one of the equatorial constellations visible in both hemispheres of Earth. It is easily recognized by the three stars forming a straight line: Orion's Belt. The other object labelled here is McNeil's Nebula (center). This is a variable nebula that was discovered in January 2004. It is variable because the star at its tip varies in brightness. The nebula is the cloud of gas and dust surrounding the star. It is about 1500 light-years from Earth.

PISCES

THE FISH

ABOVE: Pisces, the twelfth of the twelve constellations of the zodiac, is a large but inconspicuous constellation. It appears as a fish, or two fishes, in several ancient cultures. In Greco-Roman mythology, the two fish represent Venus and her son Cupid, plunging into the Euphrates when the monster Typhon attacked them. They became the fish whose images were raised into the sky.

SAGITTARIUS

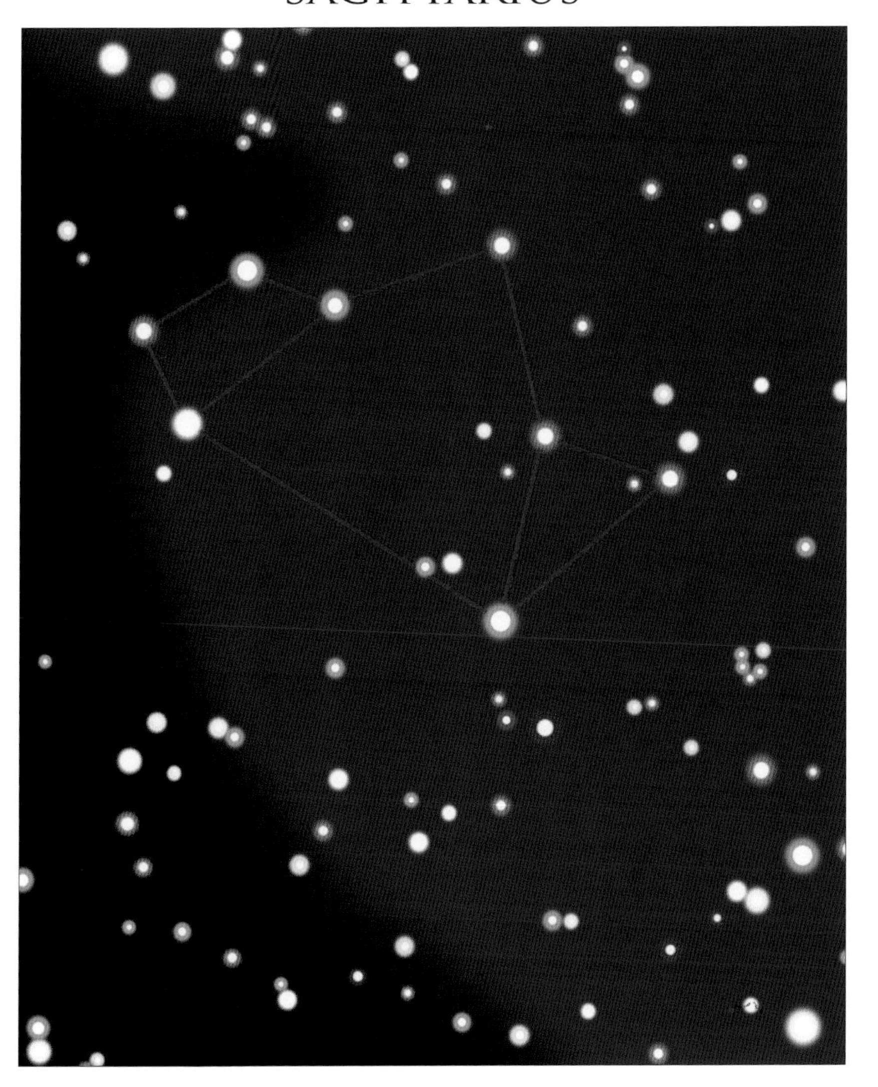

THE ARCHER

ABOVE: Sagittarius lies between Scorpius and Capricornus. It is a large southern constellation which contains the star clouds of the central Milky Way (gray), lying in the direction of the center of our galaxy. Sagittarius is depicted as a centaur holding a bow and arrow which is pointed at the star Antares, the heart of Scorpius (the scorpion). In the past, the constellation has been depicted as a human archer, a bow and arrow and, by the ancient Egyptians, as a lion.

CAPRICORNUS

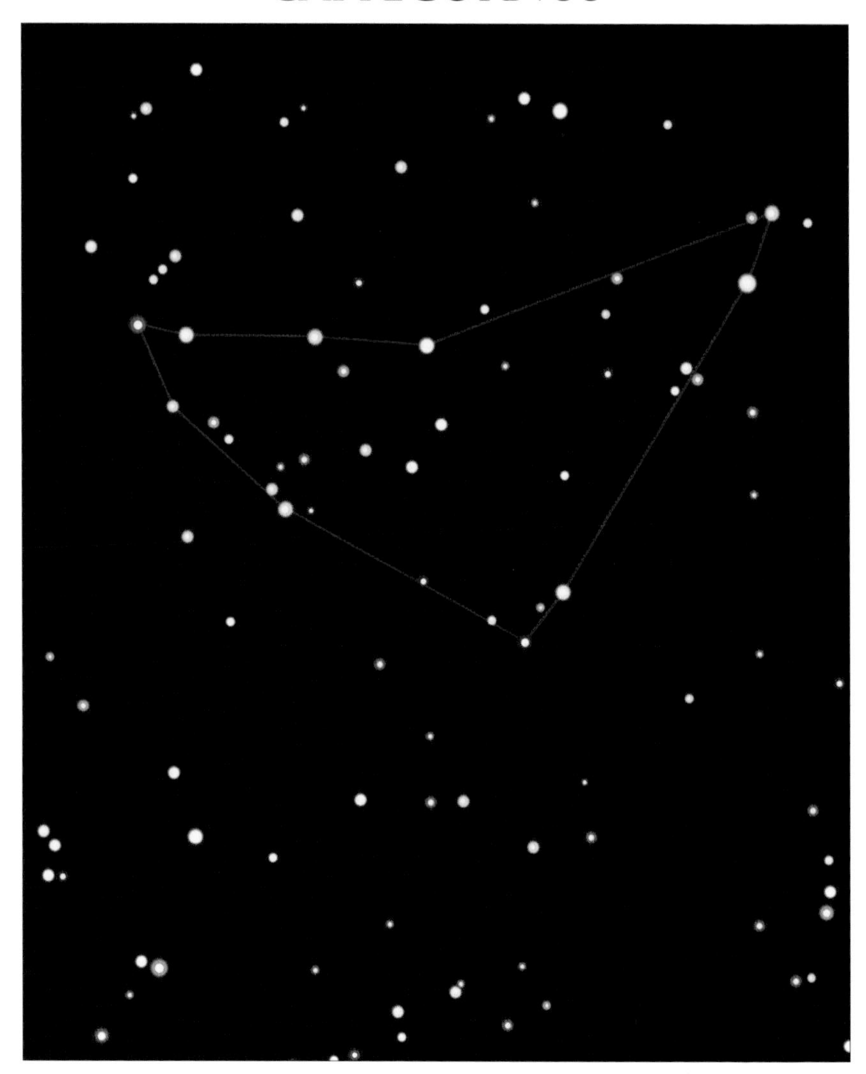

SEA GOAT

ABOVE: Capricornus lies just south of the celestial equator, but is still visible to northern observers during early evenings in the the fall. It is one of the twelve zodiacal constellations, lying between Aquarius and Sagittarius. In ancient times the whole area surrounding Aquarius was associated with water or rain.

SCORPIUS

THE SCORPION

ABOVE: Scorpius lies between Sagittarius and Libra. It is an important summer constellation for northern hemisphere observers, containing many bright stars, and the star clouds, nebulae and dust lanes of the Milky Way. In mythology, the scorpion was the creature whose sting killed Orion, so the gods placed these constellations on opposite sides of the sky.

TAURUS

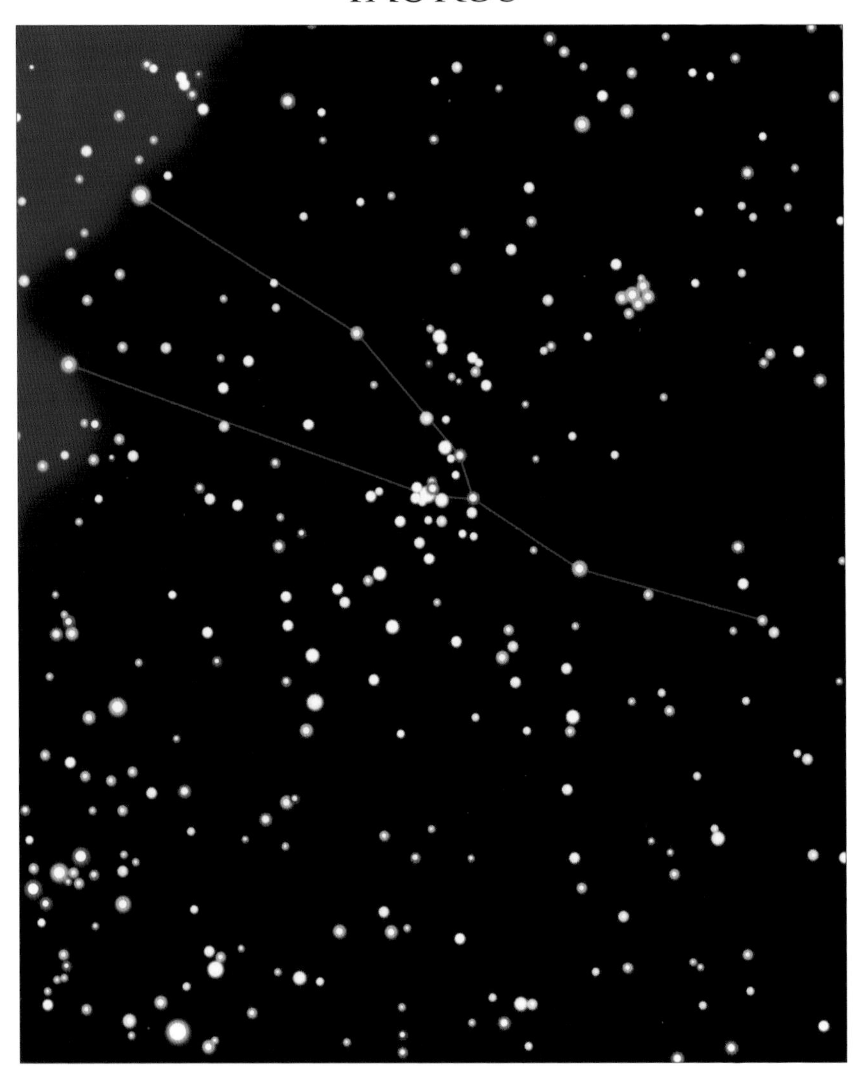

THE BULL

ABOVE: Taurus is the second of the twelve constellations or signs of the zodiac. The most distinctive feature of Taurus is the v-shape formed by the Hyades star cluster at the base of the constellation's "horns." Taurus is a very ancient constellation, possibly deriving from bull worship in early Mediterranean civilizations. In the mythology of ancient Greece, Taurus was the bull which carried Europa, only to be revealed as Zeus in disguise.

VIRGO

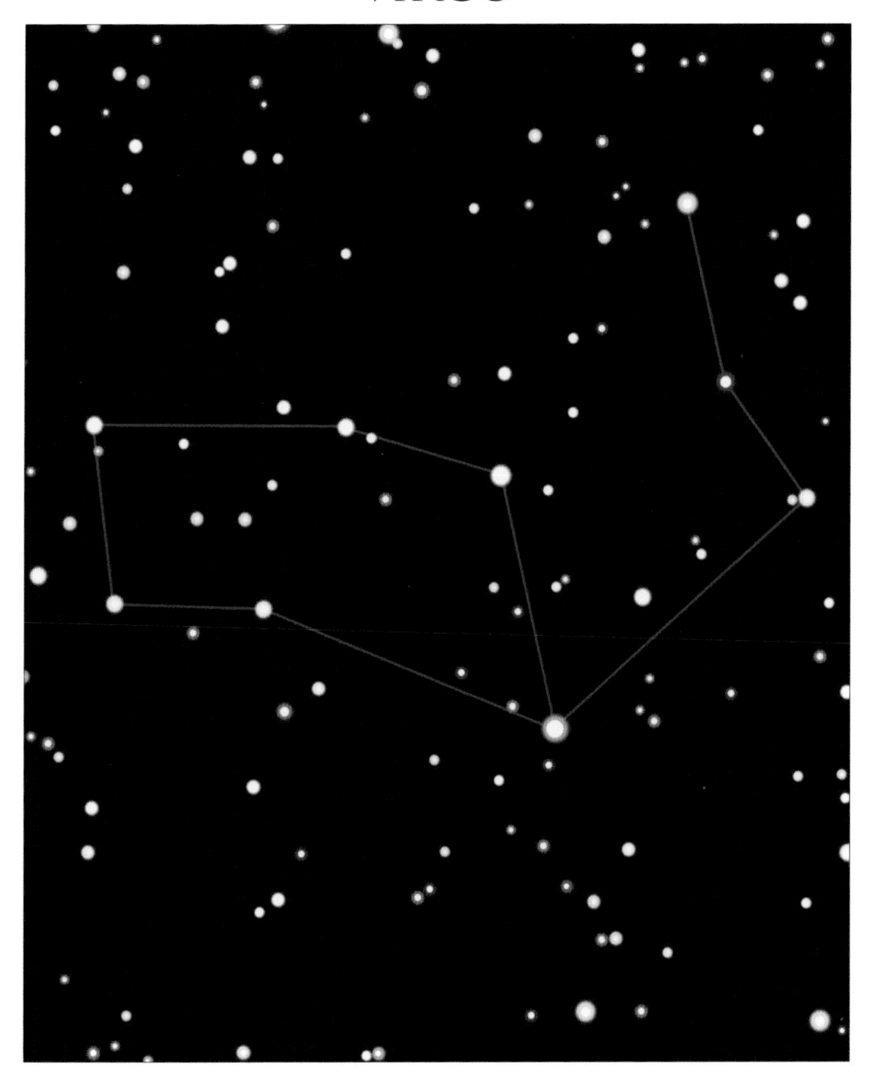

THE VIRGIN

ABOVE: The virgin is an old and astrologically important constellation associated with a mother goddess. She was Kanya, mother of Krishna, in India; Ishtar in Babylonia; Isis in Egypt. In Greece and Rome, she was Astraea, daughter of Zeus and Themis.

LIBRA

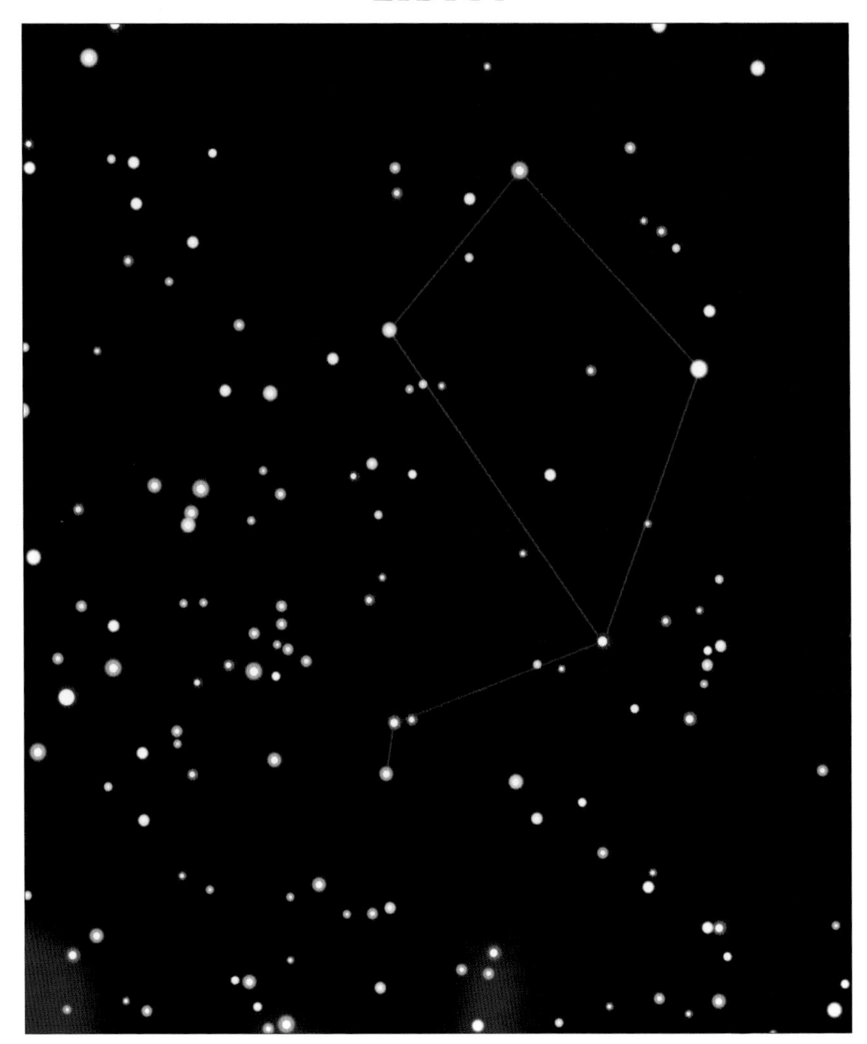

THE SCALES OF BALANCE

ABOVE: Libra is a relatively inconspicuous constellation in the southern hemisphere. Libra was apparently a creation of Roman times; before that it was joined with Scorpius in a double constellation called Scorpius cum chelae, the Scorpion with Claws. Libra was known as the balance beam in India and Middle Eastern cultures before Roman times.

NORTH CELESTIAL POLE

ABOVE: Optical image of the stars around the north celestial pole. This is marked by the pole star Polaris (Alpha Ursae Minoris, upper center) in the constellation Ursa Minor, the Little Bear. At far right, three of the bright stars in the Plough (part of Ursa Major, the Great Bear) are seen. Surrounding Ursa Minor is the faint constellation Draco, the dragon. The head of the dragon is formed by the quadrilateral of stars at lower left. The brightest of these is the orange star Eltanin (Gamma Draconis), which is just to the left of the white star Rastaban (Beta Draconis), a name meaning "head of the snake."

URSA MAJOR

THE GREAT BEAR

ABOVE: The northern constellation of Ursa Major, the Great Bear. The upper left part is also known as the Plough or Big Dipper. This is the third largest constellation in the sky. The Plough forms the back and tail of the bear, with its head and legs being represented by fainter stars at upper right and lower frame respectively.

CANIS MAJOR

THE GREAT DOG

ABOVE: At upper right is Sirius (Alpha Canis Majoris), the brightest star in the sky. It is a blue-white star which lies only 8.6 light-years from Earth. At bottom right is the second-brightest star in the sky, Canopus (Alpha Carinae), in the constellation Carina, the keel. Canopus lies around 72 light-years from Earth. Canis Major is a winter constellation for northern hemisphere observers.

CANCER

THE CRAB

ABOVE: Cancer is a relatively inconspicuous constellation. In Greek mythology, Cancer was associated with Hercules. The crab was sent by the goddess Hera to bite the hero's foot as he was struggling with the multi-headed Hydra. Hercules killed the crab by treading on it, but Hera placed its image in the heavens.

PEGASUS

THE WINGED HORSE

ABOVE: The body of the horse is formed from the square of stars at lower left. Clockwise from the orange star just left of center (Scheat, or Beta Pegasi), these are Markab (Alpha Pegasi), Algenib (Gamma Pegasi) and Alpheratz (Alpha Andromedae). Alpheratz is not actually part of Pegasus, instead lying in neighboring Andromeda. The horse's nose is Enif (Epsilon Pegasi), at lower right. The bright star at upper right is Deneb (Alpha Cygni) in the constellation Cygnus (the swan). To its right is the red North America nebula (NGC 7000).

NEBULAE

INTERSTELLAR CLOUDS

ABOVE: The Orion nebula can be seen with the naked eye as a fuzzy patch in the constellation Orion.

The term nebula, meaning "mist" in Latin, was once applied to any fuzzy patch of light in the night sky. In Charles Messier's directory of nebulae and star clusters, he identified 110 objects. Many of the objects originally thought to be nebulae were later discovered to be galaxies outside our own. However, Messier had cataloged a number of genuine nebulae, including the Crab nebula, which was his Messier Object number 1. He also indexed, among several others, the Eagle, Triffid, Omega, and Orion nebula. Nebulae are now known to be interstellar clouds of gas and dust within our own galaxy. This interstellar material is not spread evenly throughout the galaxy; instead vast areas dense in gas and dust can be found, having clumped together as a result of gravity.

STELLAR FACTORIES

ABOVE: Triffid nebula (M20, NGC 6514), Spitzer Space Telescope infrared image. This nebula is located 5400 light-years away in the constellation Sagittarius. The infrared telescope enabled scientists to view the star formation within the nebula. Within each of the four red dust clouds, comprising the lower object, there are embryonic stars. Because particles falling towards these stars get hotter with their increased energy, Spitzer can see them. It is believed that our sun was formed in such a region, but its small size led to it being gravitationally expelled.

Nebulae are factories of star formation. Our own planetary system is thought to have formed from one, the Solar nebula. There are two different types of interstellar cloud, H I regions and molecular clouds (H II regions). Interstellar clouds are usually exceptionally cold, almost minus 3000 Celsius, with the result that the atoms join together to form molecules and the cloud is known as a molecular cloud or H II region. The main constituent is molecular hydrogen, but water, ammonia and hydrocarbons are also present. Under different conditions, molecular clouds give rise to three different types of nebulae: emission nebulae and reflection nebulae, which are both optically visible, and dark nebulae which can sometimes be detected visibly if the cloud obscures our view of light shining behind it.

Regions of space less dense in interstellar matter are called H I (Hydrogen One) regions. They are composed of atomic hydrogen and are not luminous – they can only be detected by the specific 21cm radio wave that the phenomenon emits. Another way of detecting H I regions is that when they come into contact with an emission nebula they cause greater luminosity of that nebula.

EMISSION NEBULA

Interstellar gas is usually invisible and obscures our view of far-off objects. However, when the interstellar gas is near a powerful star or a tight cluster of stars, it is illuminated to form spectacular emission nebulae. For one star to illuminate the nebula, it needs to be an exceptionally hot star, which emits most of its energy as ultraviolet radiation. Photons from these stars ionize the interstellar gases, and the electrons stripped away during the ionization process try to recombine with the atomic nuclei. This process can be detected on the visible portion of the spectrum, making nebulae among the most striking sights in the galaxy.

Emission nebulae are very often reddish in color, because of the position in the visible spectrum at which hydrogen manifests itself during this process. The reason for the predominance of hydrogen in emission nebulae is not only that hydrogen is simply the most abundant element in the universe but also because less energy is required to ionize hydrogen than other elements. When the ionizing stars are so powerful that they are able to ionize other elements as well, this interrupts the uniform red with other colors.

There are dark patches within emission nebulae where interstellar dust obscures our optical view of parts of the nebula. These areas need to be charted using radio waves, which have longer wavelengths, unobstructed by the interstellar medium.

OPPOSITE: Hubble Space Telescope image showing dark pillars of dense molecular hydrogen and dust in the Eagle nebula (M16). Ultraviolet light from young stars (out of frame) evaporates gas from the one-light-year-long pillars, creating the blue halo-like effect. The small protrusions on the pillars contain globules of even denser gas which are embryonic stars; these have been dubbed Evaporating Gaseous Globules, or EGGs. The evaporation of the pillar limits the amount of gas and dust which these embryonic stars can gather. The Eagle nebula is about 7000 light-years from Earth.

REFLECTION NEBULA

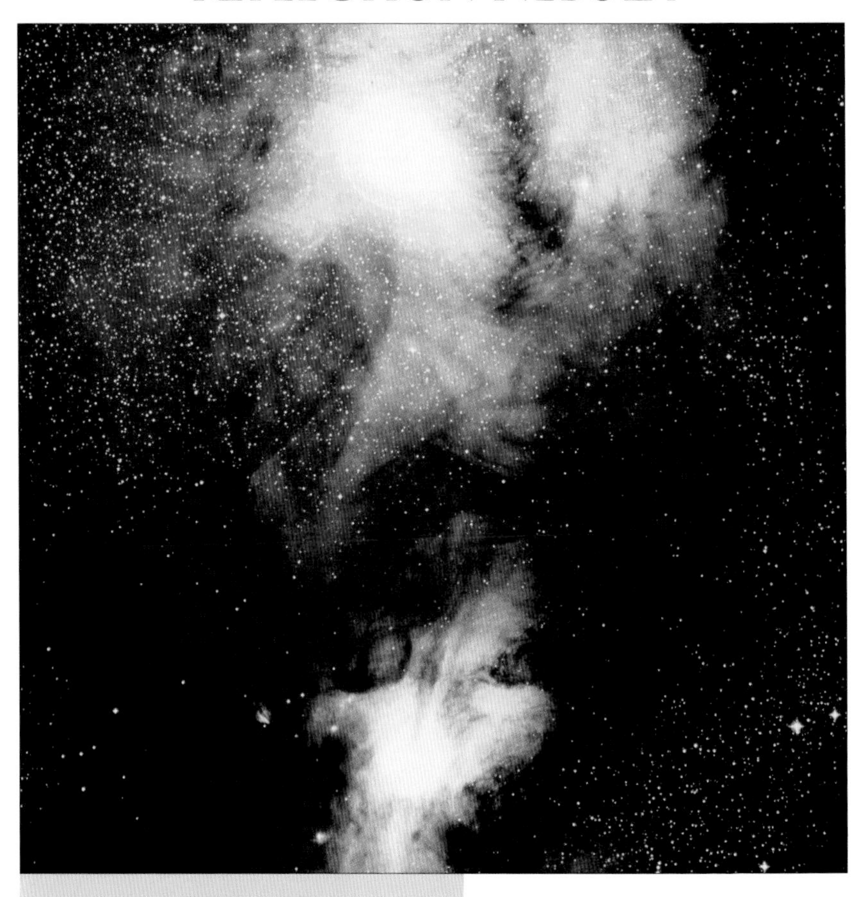

ABOVE: Rho Ophiuchi nebula. True-color optical image of nebulosity (IC 4604) surrounding the star Rho Ophiuchi (upper center) in the constellation Ophiuchus. This is an excellent example of a reflection nebula. These clouds of gas and dust do not have any light of their own, but reflect the light of nearby stars. They typically appear blue as blue light is scattered more than red light. Dark dust clouds can be seen blocking the light of background stars at lower left. The reddish star at lower right is Omicron Scorpii, in Scorpius. This image was produced by digitally combining photographs taken by the UK Schmidt Telescope in blue and red light.

When the nearby star is not hot enough to cause ionization of a nearby molecular cloud, the light from the star is instead reflected and scattered by the cloud, making it optically visible. Scattering favors the shorter wavelength, blue-colored portion of the visible spectrum meaning that most reflection nebulae appear blue.

DARK NEBULA

Dark nebulae occur when molecular clouds are not lit up by a nearby star; the particles in the cloud cannot be detected optically. Usually these nebulae go undetected against the dark backdrop of space, but sometimes they block out the view of a cluster of stars or an emission nebula, with the consequence that their impressive silhouettes can clearly be seen. The best views of dark nebulae are in the plane of the Milky Way. The thick profile of stars is frequently broken up by dark patches which are in fact dark nebulae.

ABOVE: Horsehead nebula. True-color optical image of the Horsehead nebula in the constellation Orion. North is to the left. The horsehead shape is formed by the intrusion of the dark nebula B33 into the bright emission nebula IC 434. The emission nebula shines red due to the ionization of its hydrogen gas by radiation from hot young stars embedded in it.

MILKY WAY

OUR GALAXY

The Milky Way is the name given to our own galaxy because, as it is seen from Earth, it appears to be a milky-colored brushstroke sweeping across the night sky. The fact that this whitish band is actually millions of stars was not discovered until Galileo became the first person known to observe the phenomenon through his telescope. Today, the Milky Way is thought to contain over 200 billion stars, and an immeasurable amount of interstellar gas and dust. Our solar system is found on the inner rim of the Orion Arm, a minor arm found towards the edge of our galaxy. In the eighteenth century William Herschel stated that the Earth was located in the middle of the disc because the stars seemed similar in abundance in both directions. However, this is, in fact, an optical illusion because light from stars is absorbed by interstellar dust and gas, which meant that our view is equally restricted in both directions, giving the impression that we were in the middle of the disc. It was not until 1918 that Harlow Shapley worked out the Earth's position within our galaxy by observing the distances of globular clusters.

From its position on the outskirts of the Milky Way, it takes the solar system 220 million years to complete one revolution of the core. It is impossible to know exactly what the Milky Way looks like because we do not have an objective viewpoint of our galaxy. Nevertheless, observations of other galaxies, as well as the subjective view from Earth and conscientious telescopic studies, have shown the Milky Way to be a large barred-spiral galaxy. At its center is the galactic bulge, a spheroid-shaped body of stars encasing the galactic core. Surrounding the bulge is a much flatter area of spiral arms, named the galactic disc. The disc and bulge are enveloped by a galactic halo, a thin veil of interstellar dust and dark matter broken up by globular clusters of stars. Although the amount of visible matter decreases sharply at the outer edge of the galaxy, this region of space is curiously massive. Scientists claim this mass is something called dark matter – matter which cannot be detected on the electromagnetic spectrum. The dark matter surrounding our galaxy is named the dark halo, but scientists remain unsure of its nature or origin.

RIGHT: The Milky Way from Vulpecula to Scutum.

PREVIOUS PAGE: Milky Way, optical image. Because Earth lies in one of its spiral arms, we look into the central mass of stars and see the galaxy as a band of light crossing the sky. There are numerous nebulae (pink) visible here.

ORIGINS

Much remains unknown about the formation of our galaxy; it is thought that a great cloud of gas collapsed under the weight of its own gravity between ten and fifteen billion years ago. The first stars produced by the collapsing cloud are those found in the galactic halo, and when we learn more about these stars, we will gain a greater understanding of the origins of the galaxy. The stars in the galactic bulge are similar in age to those in the galactic halo. Unlike our sun, these are Population II stars – small reddish stars that are exceptionally old – as old as the galaxy itself. The disc mostly contains Population I stars, such as our own sun, which are much younger and hotter than stars found in the bulge. Hydrogen and other gases and materials probably ended up here following the gas cloud's collapse, so the process of star formation has persisted in the galactic disc.

OBSERVING THE MILKY WAY

On a clear night, in areas devoid of light pollution, the Milky Way can be observed as a luminous band of stars cutting across the middle of the night sky. What you will actually be seeing is the galactic disc in profile. The Milky Way appears brighter near the constellations of Sagittarius and Scorpio where the stars become densely packed in the galactic bulge where the galactic core is located. However, do not expect to be able to see the core; the interstellar dust which hampers scientist from optically viewing the core, will mask it to the amateur astronomer as well.

ABOVE: The Milky Way in Scorpius and Ophiuchus. The Milky Way is the galaxy, made up of over 200 billion stars and their planets, in which our solar system is positioned. It has a mass of roughly 750 billion to one trillion solar masses and a diameter of about 100,000 light-years. Taken near Copper Mountain, Colorado.

OPPOSITE: The Milky Way from Scutum to Sagittarius.

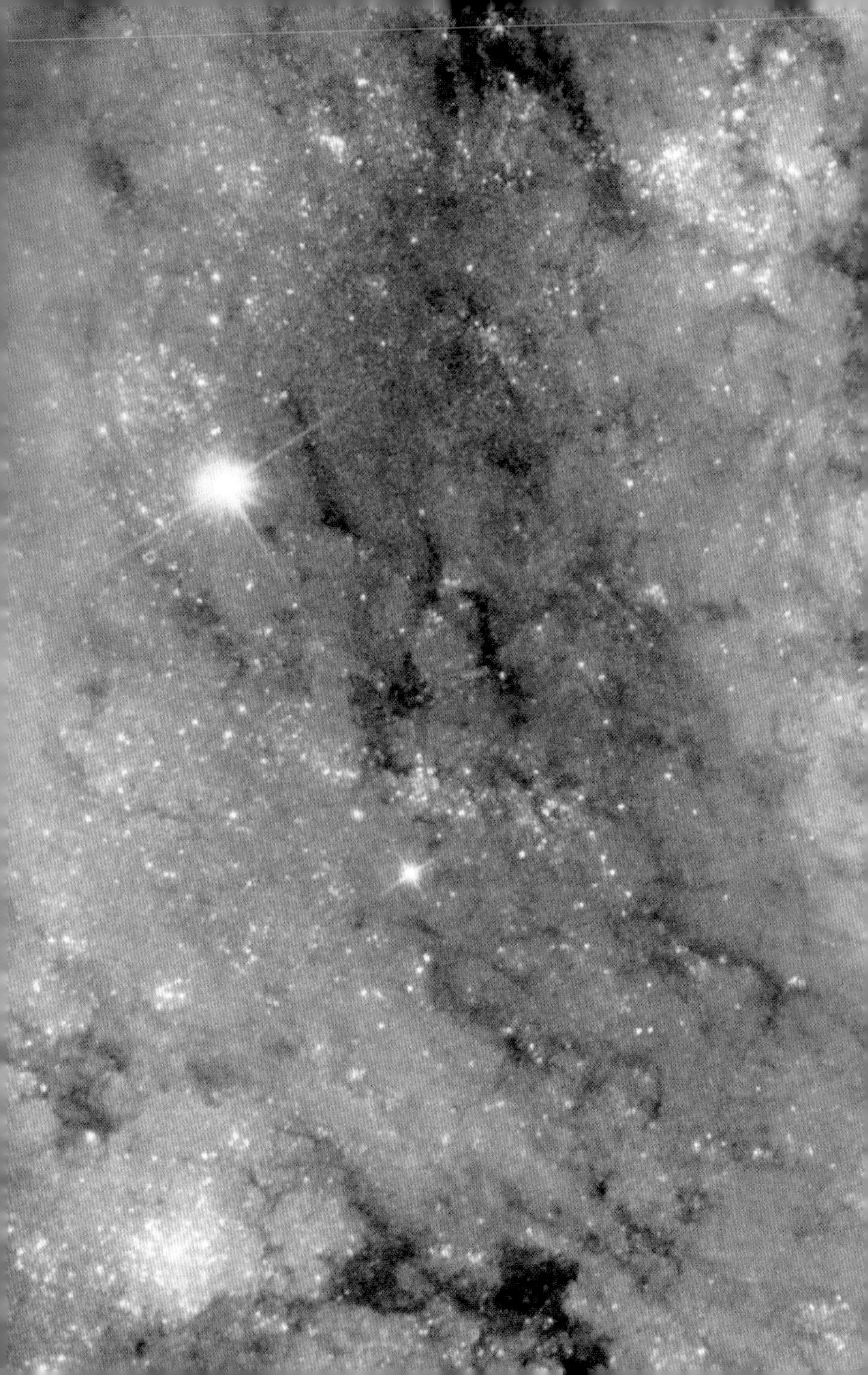

GALAXIES

EXPANDING UNIVERSE

ABOVE: Triangulum Galaxy (M33, NGC 598). Also known as the Pinwheel Galaxy, this spiral galaxy is about 2 million light-years away in the constellation Triangulum. It is relatively small, at around 30,000 light years across and is of type Sc, meaning that it has no central bar and has a loosely packed center and arms. It is part of the Local Group of galaxies that includes our Milky Way Galaxy. The galaxy contains star-forming regions that contain ionized hydrogen gas (red).

In the early 1920s, studies by the renowned American astronomer, Edwin Hubble, showed that our galaxy, the Milky Way, was not the only galaxy in the universe. For a long time it had been believed that the universe and the Milky Way were one and the same thing, but the discovery of other galaxies, outside our own, greatly expanded the size of the known universe.

Within four hundred years Earth had been consecutively demoted from its position as center of the cosmos, to being one of several planets orbiting a star, which turned out to be rather unexceptional and on the periphery of a galaxy, which proved to be just one of countless numbers of galaxies.

DISCOVERY

ABOVE: Optical photograph of the Sombrero Galaxy (M104, NGC 4594), a spiral galaxy (type Sa/Sb) in the constellation of Virgo. The galaxy is seen almost edge-on and the dark stripe running across it is due to dust lying within the plane of the galaxy. The photograph was taken with the four-meter telescope at Kitt Peak National Observatory.

Other galaxies had been observed throughout history, but few had imagined that these fuzzy balls of light did not emanate from within our own galaxy. Immanuel Kant suggested in the mid-eighteenth century that these phenomena might be a group of stars clustered together by gravity, and that what was being observed might be galaxies separate to our own. Nevertheless, these galaxies were named spiral nebulae and were commonly assumed to be within the Milky Way. During the nineteenth century, increasing evidence was gathered to support Kant's theory, but it was not until Hubble provided reliable proof to show how far away these nebulae were, that it became widely understood that these were galaxies in their own right. To measure their distances, Hubble had used the idea of a standard candle; he used the known luminosity of Cepheid stars in these nebulae as a constant and compared it to their observed brightness as a variable in order to work out their distances from Earth.

SPIRAL GALAXIES

ABOVE: Spiral galaxy NGC 6946, Gemini North telescope image. This is a starburst galaxy, one in which stars are born and die at a much greater rate than normal. This can be seen by the abundance of starbirth regions (red) in the galaxy's spiral arms. NGC 6946 was also home to eight supernovae between 1917 and 2004, making it the most prolific source of supernovae in that period. The last supernova in our Milky Way was in 1604. This galaxy lies between 10-20 million light-years away on the border between the constellations Cygnus and Cepheus. Image taken on 12 August 2004.

Many of the known galaxies are spiral galaxies; they have a bulge at the center surrounded by a flatter disc, separated into spiral arms. The stars in the disc are usually younger stars called Population I stars, while the stars in the bulge are older, Population II stars. It was initially thought that the spiral arms were caused by differential rotation – the stars in the disc closest to the bulge orbiting the galactic core at a faster rate than the stars at the outskirts, but this theory was later dismissed, as this process would quickly result in the spiral arms wrapping themselves around the bulge. Instead, it is now thought that spiral density waves might be giving the galactic disc its spiral shape. A wave would not be affected by the differential rotation, explaining why the arms of spiral galaxies did not wrap themselves around their galactic bulge a long time ago.

ELLIPTICAL GALAXIES

Elliptical galaxies do not have the spiral arms, bright bulges or the flat discs of spiral galaxies, instead they appear to be a block of stars, brighter towards the galaxy's center and slowly fading at the outskirts. Hubble initially believed that elliptical galaxies might evolve into spiral galaxies, but this is not actually the case. Stars in elliptical galaxies are much older than those found in the disc of spiral galaxies and are of similar ages to the stars found in the galactic bulge or halo. These are mainly Population II stars, which are much older than stars like our own sun; additionally, there does not appear to be any star formation occurring in elliptical galaxies, despite the fact that recent observations of elliptical galaxies have revealed younger star clusters inside some elliptical galaxies. This phenomenon seems to be best explained by a galaxy merger in which the gravitational pull of a larger galaxy draws in smaller galaxies and globular clusters.

ABOVE: Hubble Space Telescope image of the center of the active galaxy Centaurus A, which is thought to house a massive black hole. Centaurus A is a large elliptical galaxy which collided with a smaller spiral galaxy millions of years ago, sparking a huge burst of starbirth activity. Large areas of young blue stars are seen on either side of the dark lane of dust. This dust is thought to be the remnant of the smaller galaxy which Centaurus A swallowed. Centaurus A is also a powerful emitter of radio waves powered by its black hole. It lies 10 million light-years from Earth in the constellation Centaurus.

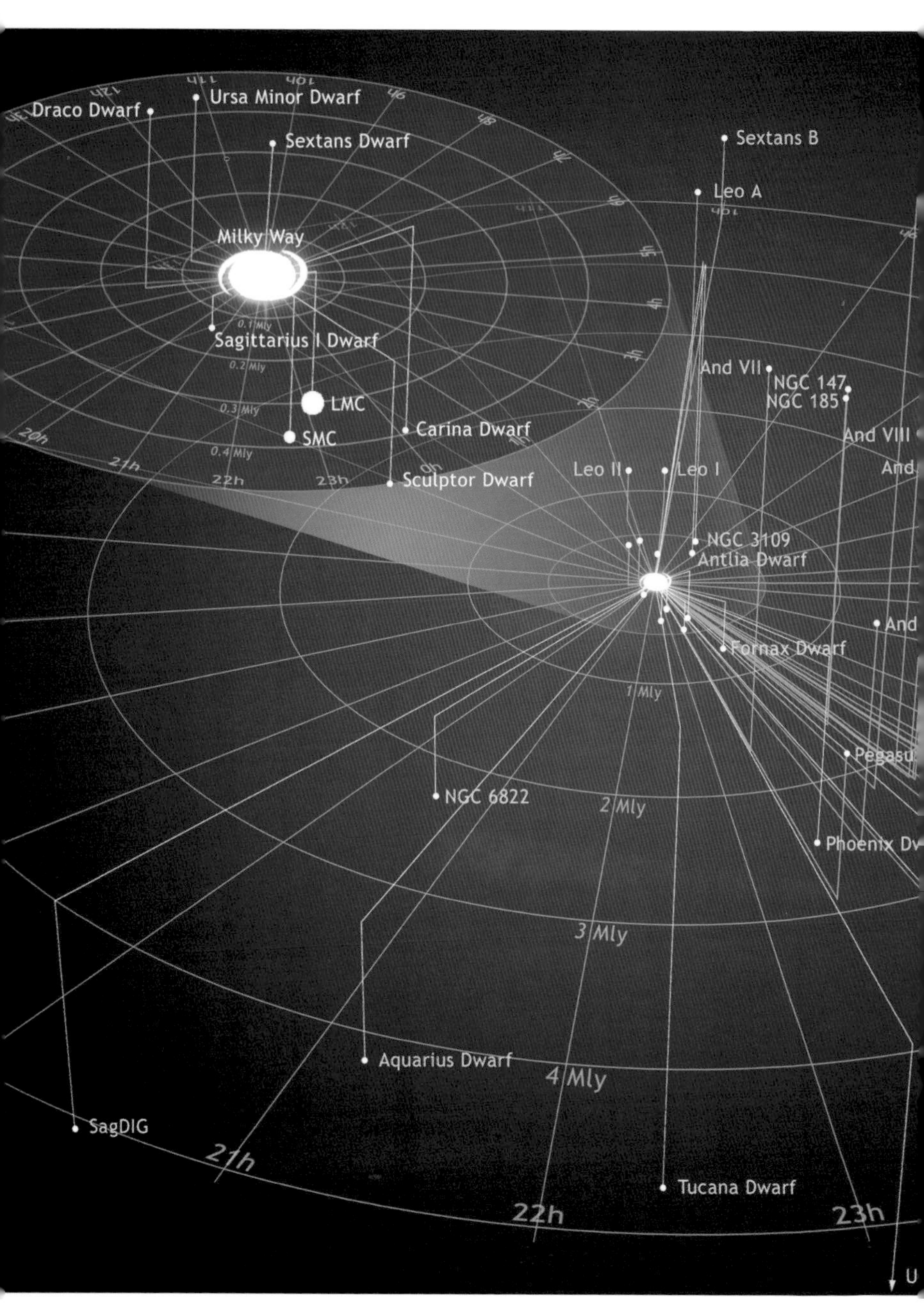
Draco Dwarf
Ursa Minor Dwarf
Sextans Dwarf
Sextans B
Leo A
Milky Way
Sagittarius I Dwarf
0.1 Mly
0.2 Mly
0.3 Mly
0.4 Mly
LMC
SMC
Carina Dwarf
Sculptor Dwarf
20h
21h
22h
23h
And VII
NGC 147
NGC 185
And VIII
Leo II
Leo I
NGC 3109
Antlia Dwarf
Fornax Dwarf
1 Mly
NGC 6822
2 Mly
Phoenix Dv
3 Mly
Aquarius Dwarf
4 Mly
SagDIG
21h
Tucana Dwarf
22h
23h

LOCAL GALAXIES

Most galaxies in the observable universe are moving away from us as a result of the great explosion which created the universe, the Big Bang. This movement has been proved because stars traveling away from us emit a longer, redder wavelength; this phenomenon, known as "red shift," indicates that an object is moving away from the observer. However, the closest galaxies to us emit a shorter wavelength, which appears bluer on the visible spectrum, which indicates that they are moving towards us. This occurs because the force of gravity has overcome the force of the explosion, a process which results in the creation of galaxy clusters. The galaxy cluster to which the Milky Way belongs is called the Local Group, which consists of around thirty galaxies. The Milky Way is the second largest galaxy in the Local Group, the largest being Andromeda. Together these two galaxies dominate the cluster; twelve smaller galaxies are satellites of the Milky Way and the remainder form the entourage of the Andromeda galaxy. The Local Group's center of gravity is located in between Andromeda and the Milky Way. The third largest galaxy, the Triangulum Galaxy, is the only other spiral galaxy in the Local Group. Although it is a satellite of Andromeda, scientists are trying to work out whether a small, irregular galaxy named LGS 3 is actually a satellite of Triangulum and not Andromeda as previously thought. The fourth largest galaxy, the Large Magellanic Cloud, together with its near neighbor, the Small Magellanic Cloud, are the closest galaxies to the Milky Way and can be observed with the naked eye from the southern hemisphere in areas unaffected by light pollution. In the northern hemisphere, neither galaxy can be observed, which means they were not officially discovered until Magellan, their namesake, embarked upon his journey around the world in the early sixteenth century.

LEFT: Local Group galaxy cluster, computer graphic. Some 40 galaxies of the central area of the Local Group are shown relative to our spiral Milky Way Galaxy (center). The concentric circles are 1 million light-years apart, and lines show the positions of galaxies above (green) or below (orange) the celestial equator. The radial lines show the directions in the sky. An inset (upper left) shows the galaxies within 500,000 light-years of the Milky Way. The largest galaxies are the Milky Way and Andromeda (M31, upper right) and M33 (right of Andromeda). The inset includes the two satellite galaxies of the Milky Way, the Large and Small Magellanic Clouds (lower center).

IRREGULAR AND LENTICULAR

RIGHT: Optical image of the Whirlpool Galaxy (M51, NGC 5194) and its companion NGC 5195. These interacting galaxies are about 20 million light-years away in the constellation Canes Venatici. M51 has a diameter of around 65,000 light-years and a mass of about 50,000 million solar masses, about half the mass and size of the Milky Way. The colors show the age of the stars in the galaxies. The nuclei contain mainly older, yellower stars, while the spiral arms have younger, hot, blue stars. The galaxies are linked by a thin bridge of gas and dust which NGC 5195 is pulling from the larger galaxy by gravitational attraction.

Irregular galaxies do not fit into either elliptical or spiral categories. They do not seem to be organized in any particular fashion. Irregular galaxies comprise younger, Population I stars and still undergo star formation. Irregular galaxies are the youngest type of galaxy. It is thought that in time these irregular galaxies will evolve into either spiral or elliptical galaxies – meaning that the life cycle of galaxies is either from irregular through spiral to elliptical or directly from irregular to elliptical.

When galaxies bear resemblance to both elliptical and spiral galaxies, they are called lenticular galaxies. These lens-like galaxies have a similar shape and age as elliptical galaxies but, like spiral galaxies, are surrounded by a disc, although unlike spiral galaxies they do not have arms.

ANDROMEDA

Andromeda is the major galaxy of the Local Group. It is one and a half times the mass of the Milky Way and it is currently thought to be about 2.9 million light-years away, making it the most distant object most people would be able to find with the naked eye. The Andromeda Galaxy is one of 110 objects visible in the night sky in the northern hemisphere which were cataloged by the French astronomer, Charles Messier. Cataloged as the 31st Messier Object, Andromeda will often be referred to as M31. In 1924, M31 became the first place known to exist outside our own Milky Way, as it was Cepheid stars in the Andromeda "nebula" that first indicated to Hubble that we were dealing with a foreign galaxy. The Andromeda Galaxy and the Milky Way are moving towards one another at an incredible pace; it is thought that the two galaxies will impact one another in three billion years' time, which will eventually lead to the creation of one giant elliptical galaxy.

ABOVE: Optical image of the Andromeda Galaxy, the nearest major galaxy to our own Milky Way. It is the largest galaxy in the Local Group, being 150,000 light-years in diameter. It lies 2.9 million light-years away. It is classed as a spiral galaxy (Sb). The Andromeda Galaxy has two smaller companion galaxies: M32 (NGC 221, top right corner) and M110 (NGC 205, lower center). Photographed by the 0.9-meter telescope at Kitt Peak National Observatory, Arizona, USA.

THE UNIVERSE

COSMOLOGY

STEADY STATE THEORY

In 1948, Fred Hoyle, Thomas Gold and Hermann Bondi proposed that the universe is eternal and infinite in time and size – it did not have a beginning and it would not have an end. While this theory appears at first glance to suggest a static universe, these three scientists did not refute the new evidence of an expanding universe, but reconciled it with their idea of a steady state of affairs. Arguing that the universe does not change in its appearance over time, they claimed that when galaxies recede from the Milky Way and disappear from view, new matter creates new galaxies so that the density of the universe remains constant; thus the observable universe appears unchanged.

ABOVE: Hubble Space Telescope view of lensed galaxies (arcs) seen through the galaxy cluster Abell 1689. Abell 1689 lies around 2.2 billion light-years from Earth. It is so massive that it bends the path of the light from galaxies even further away behind it, acting like an enormous lens in space. This means it can be used to study more distant galaxies than would be possible to observe without the cluster. The fine arcs are the distorted images of galaxies from up to 13 billion light-years away, seen as they were when the universe was less than a billion years old.

PREVIOUS PAGES: Hubble Space Telescope deep-view image of several thousand never-before seen extremely distant galaxies. This is the deepest view yet into the universe. The most distant of these galaxies are around 12 billion light years away, meaning their light has travel across about three-quarters of the known universe to reach Earth. These galaxies lie in the southern constellation Tucana.

THE BIG BANG

Even before the Steady State theory had been proposed, in 1927 a Belgian priest, George Lemaitre, suggested that the universe had begun with the explosion of a "primeval atom." This later became known as the Big Bang, a name which had initially been a derogatory term, coined by Fred Hoyle in order to make the whole idea sound absurd. Nevertheless, this notion of a single atom from which the universe germinated gained supporters such as George Gamow, who developed the idea much further than Lemaitre, and is often credited as being the father of the theory. The idea is that if the universe is expanding and the galaxies are moving away from one another, (as had been shown by red shift), then at some stage in the history of the universe, there must have been a point of origin, which was perhaps a gravitational singularity. At this point of origin, there must have been a great explosion, and ever since that moment, the universe has been expanding and cooling.

So, while the Big Bang postulation had been around for a number of years, during the 1950s the Steady State hypothesis was the more popularly accepted theory to explain the creation of the universe; the concept of nothingness before the Big Bang was difficult to fathom and the whole idea seemed too far-fetched.

The Steady State theory enjoyed popularity during the 1950s but it soon came up against a number of challenges which swayed opinion in favor of the "Big Bang" as the more likely explanation of the evolution of the universe. One problem was the discovery of very young galaxies called radio galaxies with their strong radio emissions. The Steady State theory holds that young galaxies are formed throughout the universe while the Big Bang theory maintains that there are no young galaxies because all have aged since the time of the Big Bang, but the furthest galaxies would still appear young because it has taken billions of years for light and other radiation to reach us. Radio galaxies were discovered to be very far from the Earth, seemingly a vindication of the Big Bang.

This blow to the Steady State theory was compounded in 1963 by the discovery of powerful radio sources, later named quasars (quasi-stellar radio objects). Soon after, a number of quasars were discovered emitting strong sources of various types of electromagnetic radiation, not just radio waves. These sources were determined to be billions of light-years from Earth; like radio waves, quasars undermined the Steady State theory which would suggest that galactic phenomena such as quasars should be scattered throughout the galaxy, not just on the farthest reaches of visible space. Quasars were easily reconciled to the Big Bang theory; they would have been part of the universe in its very early stages, and could only be visible at such vast distances because it took so long for light and other radiation from them to reach us.

Although the discovery of distant quasars and radio galaxies discredited the Steady State theory, it did not actively support the Big Bang theory. Advocates of the latter would need to find hard evidence to back up their claim. They believed that after a huge explosion the universe would have emitted radiation, which would have spread evenly

SUPPORTING THE BIG BANG

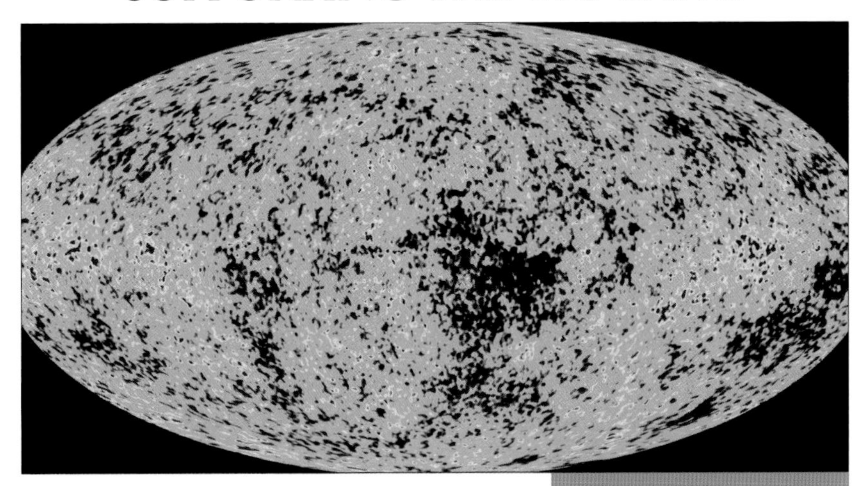

ABOVE: Cosmic microwave background. Whole sky image of the cosmic microwave background made by the MAP (Microwave Anisotropy Probe) spacecraft. This map indicates that the age of the universe is around 13.7 billion years. The data also reveal that the universe is expanding at 71 kilometers per second per megaparsec (1 Mpc = 3262 light-years). The colors reveal variations in the temperature of the universe in all directions. This correlates to the density of material at the time when the universe became transparent to radiation, about 380,000 years after its creation. The denser regions (red, yellow) formed the seeds of galaxies and other structures. Data obtained in 2003.

throughout space. Essentially, if there had been a Big Bang, this radiation should be everywhere in space and if the existence of this radiation could be detected, it could support the Big Bang theory.

In 1965, Arno Penzias and Robert Wilson working for Bell Telephone Laboratories in New Jersey were pioneering the development of satellite technology for telephones. To get the signal from the satellite, they needed to remove all interference. However, every time they tried to link up with the satellite, a faint, constant background interference persisted in blocking their attempts. It did not seem to be coming from one specific location, its source was multi-directional, issuing from everywhere in space. Accidentally, they had discovered the radiation the Big Bang theorists had been searching for; it was a microwave, later named cosmic background radiation. This discovery provided the Big Bang theory with some vital supporting evidence, which aided its progress in rapidly replacing the Steady State theory as the main model for describing the evolution of the universe.

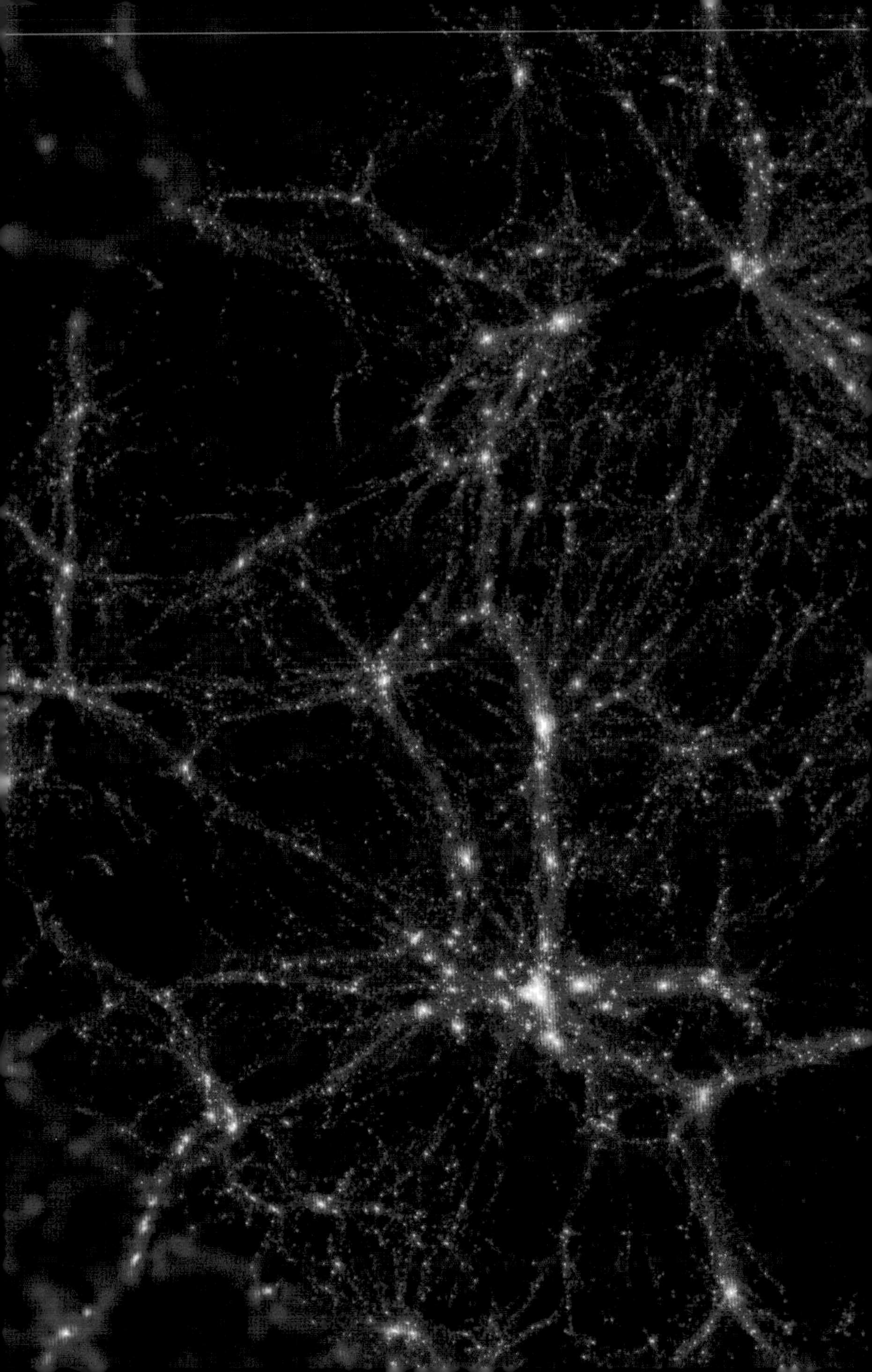

THE BIG CRUNCH

Scientists are faced with the question of whether or not the universe will keep on expanding, or whether it will begin contracting at some point in the future. Some of the galaxies closest to the Earth, such as Andromeda, emit a blue shift, rather than a red shift, which implies that they are moving towards us because, when an object approaches an observer, it emits a shorter, bluer wave. The reason these galaxies are moving towards us is because gravity is sufficiently strong to overcome the recession of nearby galaxies.

Some Big Bang theorists have suggested that this blue shift indicates that the universe will end in a Big Crunch, when the expansion of the universe is halted and reversed until all matter in the universe contracts back into the gravitational singularity from which it came – at which point time would come to an end.

For the Big Crunch to occur the universe needs a certain density of matter and energy to reverse the expansion; this is called the "critical density." The Big Bang theory holds that matter and energy are not created, the universe was as dense as it was ever going to get at the time of the Big Bang. Cosmologists, therefore, are trying to calculate the density of the universe to see whether it is above or below the critical density.

Recent calculations suggest that the universe might be less dense than the critical density, which would result in the universe expanding for ever. However, scientists have difficulty factoring in dark matter and energy when calculating the density of the universe; thus unbounded expansion or the Big Crunch both remain distinct possibilities.

There are some Big Crunch theorists who think that the whole process might repeat itself as soon as the universe has contracted, in other words there may be a new Big Bang. If that is the case, then this process could have happened millions of times before and our universe is just one of a sequence of many universes.

LEFT: Supercomputer simulation of the distribution of dark matter in the local universe. Dark matter is a form of matter that cannot be detected by telescopes as it emits no radiation. Its presence is inferred by its gravitational action on visible matter. It is thought that visible matter (such as galaxies) occupies halos of dark matter (bright dots). The halos are connected by filaments of cold, invisible dark matter (red). This simulation matches the observed distribution of galaxies well, suggesting current theories about the universe are accurate.

INDEX

A

B

C

INDEX

D

E

G

Q

R

S

All images are courtesy Science Photo Library

David P. Anderson SMU/NASA (42);
Julian Baum (200,201,202,203,206,207,208,209,210,211,212,213,216);
Wesley Bocxe (65); Chris Butler (129,160,169);
California Association For Research In Astronomy (136,142);
Celestial Image Co. (224); John Chumack (165,170,190,229,230,231);
Chris Cook (84); Tony Craddock (16); ESA/DLRFU Berlin/G. Neukum (86,96,99);
Dr Fred Espenek (30,177, 186); European Southern Observatory (194);
European Space Agency (98,135); Mark Garlick (11,12,36,48, 126,188,238);
Robert Gendler (218,220); Francois Gohier (178); H420/131 (163);
Tony & Daphne Hallas (226,234); Roger Harris (62, 214); Mehau Kulyk (19);
Larry Landolfi (78,85); M. Ledlow (37); Jerry Lodriguss (22,23176);
Jon Lomberg (205); Max Planck Ins. For Astrophysics (249);
G. Antonio Milani (80);
Mount Stromlo & Siding Spring Observatories (192,221,225);
NASA(2,18,26,28,31,33,38,40,45,46,47,49,50,52,57,58,61,66,68,70,72,73,76,77,82
89,95,100,101,102,106,108,109,110,112,114,116,117,118,120,123,127,128,130,132
133,140,141,148,150,151,152,154,156,159,162,172,174,175,187,247);
NASA/ESA/STScl
(3,90,122,124,134,139,145,158,166,167,191,222,232,237,242,245)
NASA/JPL/Cornell (4,92); Adam Nieman (166); NOAO (235,241);
NOAO/AURA/NSF (236); David Nunuk (14); Parker (244);
Planetry Visions Ltd (54,56,60); John Sanford (69);
Eckhard Slawik (44,74,79,198,204,215,217); SOHO/ESA/NASA (20,24,27);
Sheila Terry (138); Joe Tucciarone (88,193);
Ulli Steltzer, US Geological Survey (34,144,146);
Detlev Van Ravenswaay (8,32,41,94,97,104);
Jason Ware (240); Frank Zullo (10);

Acknowledgements

This book would not have been possible without the help of the following people:
Jane Benn, Mathlida Hoyle, Hayden Wood, Richard Betts, Kevin Davis,
Sarah Rickayzen, Cliff Salter, Melanie Cox.
Design by John Dunne.